Rabbit Biotechnology

Louis-Marie Houdebine • Jianglin Fan
Editors

Rabbit Biotechnology

Rabbit Genomics, Transgenesis, Cloning and Models

Springer

Editors
Louis-Marie Houdebine, Ph.D
Doctor, Institute of Research in Agronomy
Jouy en Josas
France

Jianglin Fan, MD, Ph.D
Chairman and Professor
Department of Molecular Pathology
Interdisciplinary Graduate
School of Medicine and Engineering
University of Yamanashi
Yamanashi
Japan

ISBN 978-90-481-2226-4 e-ISBN 978-90-481-2227-1
Springer Dordrecht Heidelberg London New York

Library of Congress Control Number: 2009921132

Printed on acid-free paper

Springer is part of Springer Science+Business Media (www.springer.com)

Contents

The first international transgenic rabbit conference cut held in Tsukuba, Japan (May 2005)

The 2nd international conference on rabbit biotechnology held in Jouy en Josas, France (June 2007)

Chapter 1
Introduction

Louis-Marie Houdebine and Jianglin Fan

The study of biological functions of proteins and their possible roles in the pathogenesis of human diseases requires more and more relevant animal models. Although mice including genetically modified mice offer many possibilities, other non-murine species are absolutely required in some circumstances. Rabbit is one of these species, which has been widely used in biomedical studies. This animal is genetically and physiologically closer to humans including cardiovascular system and metabolism characteristics. Rabbit is thus more appropriate than mice to study some diseases such as atherosclerosis and lipid metabolism. Because of its larger size, surgery manipulation, bleeding, and turn-over studies are much easier performed in rabbits than in mice. Furthermore, transgenic rabbits can be produced using microinjection and other methods such as lentiviral vectors. Cloning in rabbits has been proved possible, even though still laborious and time-consuming. Hopefully, functional rabbit ES cell lines will be available in the coming years. Gene deletion or knock-out in rabbits will then become possible. In the mean time, gene knock down using siRNA or micro RNA is an attractive alternative. The accomplishment of the whole rabbit genome sequencing is about to be achieved. Moreover, rabbit is being used to produce pharmaceutical proteins, including human polyclonal antibodies. Rabbit is also a significant source of meat in some countries. These biotechnology projects, although very different, are using essentially similar technical approaches. An optimal application of rabbits requires improvement of these different techniques. To exchange the information and update the advanced technology in rabbits, the first international meeting on rabbit

L.-M. Houdebine (✉)
Biologie du Développement et Reproduction, Institut National de la Recherche Agronomique 78350 Jouy en Josas, France
e-mail: louis.houdebine@jouy.infra.fr

J. Fan (✉)
Department of Molecular Pathology, Interdisciplinary Graduate School of Medicine and Engineering, University of Yamanashi, Yamanashi, Japan
e-mail: jianglin@yamanashi.ac.jp

L.-M. Houdebine and J. Fan (eds.), *Rabbit Biotechnology: Rabbit Genomics, Transgenesis, Cloning and Models*,

biotechnology was held in Tsukuba, Japan, in 2005 and followed by the second meeting in Jouy en Josas (near Paris) on June 14 and 15, 2007. The next meeting has been scheduled on June 4 and 5, 2009 in Xi'an, China. A specific biotechnology of the rabbit is therefore emerging. We strongly feel that the time has come to compile a contemporary professional book (tentatively named as Rabbit Biotechnology). This book aims at confronting the different sophisticated approaches developed independently in different academic and industrial laboratories as well as in some companies in the world.

Chapter 2
Improvement of Rabbit Production

Shuji Kitajima

Abstract In this chapter, I will introduce some of the techniques for efficient colony management and production of rabbits. The artificial insemination (AI) can be a useful technique which shows a better performance rather than natural mating. The pregnancy rate and mean litter size after AI are not different from those found after natural mating. Moreover, judging from our result, one ejaculate from one male rabbit can be sufficient to fertilize about 25 female rabbits with AI. In addition to AI, sperm freezing is also an important technique for a stable maintenance of rabbit strains for long time at low cost. These techniques can contribute to an enhancement of productivity and stability of maintenance of rabbit colonies.

Keywords Artificial insemination, Cryopreservation, Rabbit breeding, Sperm freezing

2.1 Introduction

Inbred rabbits are not commonly used, therefore, an appropriate colony management is certainly necessary for the maintenance of rabbit breeds to avoid inbreeding deleterious effects. Colony management implies a lot of rabbits and a large space for breeding and maintenance. However, it is difficult for most laboratories to find enough space for rabbit breeding and maintenance because of the relatively large scale of animal facilities and for financial reasons. Moreover it is necessary to save space and money in many cases.

Currently, in domestic animals, especially in cattle, artificial insemination (AI) using frozen sperm is widely implemented in the world. It is generally considered that this

---

S. Kitajima (✉)
Division of Biological Resources and Development, Analytical Research Center
for Experimental Sciences, Saga University 5-1-1 Nabeshima, Saga 849-8501, Japan
e-mail: kitajims@cc.saga-u.ac.jp

L.-M. Houdebine and J. Fan (eds.), *Rabbit Biotechnology: Rabbit Genomics, Transgenesis, Cloning and Models*,

technique contributes very significantly to an enhancement of productivity. In rabbits, many studies on AI have been reported. This method is very simple and applicable for systematic colony management rather than the natural mating. There are lots advantages in AI for rabbit reproduction. For instance, we can use diluted semen for AI, usually semen from one male is sufficient to fertilize more than ten female rabbits at one time. It enables rapid expansion of rabbit colony from even small number of rabbits at a start. In the research field, we sometimes need a lot of rabbits in same age at one time for experiments. By reproduction with AI, we can prepare them easily at a given time. Moreover, recent advances of reproductive science have brought us a number of advantages such as cryopreservation of sperm and embryos in liquid nitrogen, which allows us to maintain stably them for long time at low cost. The combination of AI and cryopreservation of rabbit semen can be useful techniques for the maintenance of rabbit breeds. For efficient production of rabbits, AI is a useful technique, it can contribute to an enhancement of productivity and reduction of the number of rabbits for colony maintenance. In this chapter, I introduce a technique of AI and cryopreservation of rabbit semen which are used routinely in our laboratory.

2.2 Artificial Insemination (AI)

2.2.1 Rabbit Semen Collection

Semen collection can be performed using artificial vagina (AV) (Morrell 1995). We can purchase AV from company, but we can made it easily (Morrell 1995; Naughton et al. 2003). The basic structure of the AV is assembled by the appropriate size of outer case and inner rubber which lines over the top and bottom edges of inner case. The test tube is connected to the top hole of the case, and the space between inner case and lined rubber is filled with hot water (Fig. 2.1). The most important point is to control temperature of hot water (It is better within 40–45°C at time of use) placed in the AV since rabbit ejaculates by heat stimulation. If temperature of hot water is too hot, it often induces contamination of urine into ejaculate. A decoy animal used for semen collection is enough to be a male, it is not necessary to be female. When two male rabbits are placed in same box (enough size for adult two rabbits) or decoy male rabbit is introduced in a cage of candidate rabbit for

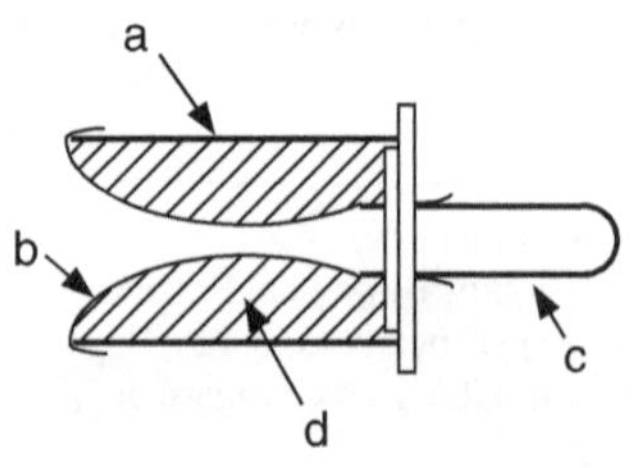

Fig. 2.1 Basic structure of artificial vagina (AV). The AV is assembled by the appropriate size of outer case (**a**) and inner rubber (**b**) which lines over the top and bottom edges of inner case. The test tube (**c**) is connected to the top hole of the case, and the space between inner case and lined rubber is filled with hot water (**d**)

semen collection, well trained rabbit will give mounting action in a few minutes. When male rabbit does the action, lead its penis into the AV immediately to collect the ejaculate (Fig. 2.2). After semen collection, volume of ejaculate should be measured. Sperm concentration and motility should also be counted immediately under microscope. The semen showing contamination by urine, extremely low sperm

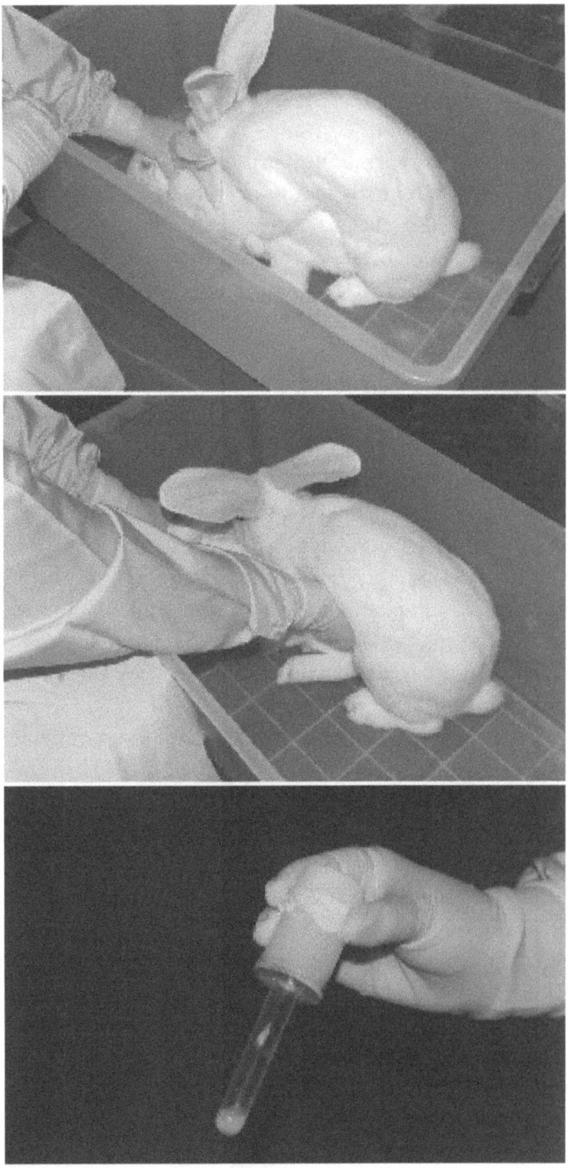

Fig. 2.2 Collection of rabbit semen. *Top*: When two male rabbits are placed in same box, well trained rabbit will give mounting action in a few minutes. *Middle*: When male rabbit does the action, lead the penis into the AV immediately to collect the ejaculate. *Bottom*: Photograph shows the collected semen by our handmade AV

concentration, a large proportion of abnormal sperm or a low motility should not be used for AI. It is generally considered that semen collection from a given rabbit should better not be performed more than twice in a week. In our laboratory, mean semen volume, sperm concentration and motility in over 5 months old Japanese White (JW) rabbits (from 5 to 24 months old, n = 202) was 0.53 ml, 600.4×10^6 spermatozoa/ml and 82.7%, respectively.

2.2.2 *Insemination*

When AI is performed immediately after semen collection, dilution of semen can be appropriate with saline. If there are several hours after semen collection until it is used for AI, semen should be diluted in an appropriate buffer such as tris citrate glucose (TCG) buffer and kept under appropriate temperature (El-Gaafary 1994; Roca et al. 2000). The sperm can maintain a normal fertility for several hours and up to a few days if it is stored in appropriate buffer and temperature (Lopez-Gatius et al. 2005; Roca et al. 2000). (Please see Section 2.5.)

For insemination, the pipettes with a 5 mm diameter of polyethylene, plastic and glass tubes can be used and it needs to be at least 10–15 cm in length. Vagina of rabbits is long and there is distance to the orifice of uterus. Therefore, the pipette should be inserted 10–15 cm into vagina to ensure appropriate delivery of sperm to female rabbits (Fig. 2.3). The glass tube bent slightly at the portion from tip 4–5 cm is easily inserted into the depth of vagina. When insemination is done by two persons, one retains rabbit by holding her back and the other should operate the glass tube. If rabbit is anesthetized by such as thiamylal sodium, which is an ultra-short acting anesthetic, it can be performed safely and surely by one person without any accident. At the time of insemination, ovulation of rabbits must have been induced by injection of 50 U of human chorionic gonadotropin (hCG).

A study on the relationship between the number of spermatozoa for AI and the pregnancy rate has been done by Viudes de Castro and Vicente (1997) who reported that the reproductive performance after inseminating 4 or 10×10^6 total sperm was not different as judged by fertility (74%) and litter size at birth (9.0 pups). On the contrary, female rabbits inseminated with 2 or 1×10^6 sperm had a significantly lower pregnancy rate: 66% and 23%, respectively (Viudes-De-Castro and Vicente 1997). Castellini and Lattaioli (1999) also reported similar results showing that the reproductive performance is independent of sperm number within a certain range. The low sperm motility in samples containing less than 4×10^6, and particularly less than 2×10^6 have a negative effect on the reproduction rate. In our data, the number of 10×10^6 spermatozoa/0.5 ml/doe was mostly efficient in JW rabbit, the pregnancy rate and mean litter size were 80% and 7.6 pups, respectively. These data are not different from those found after natural mating. It can even be considered

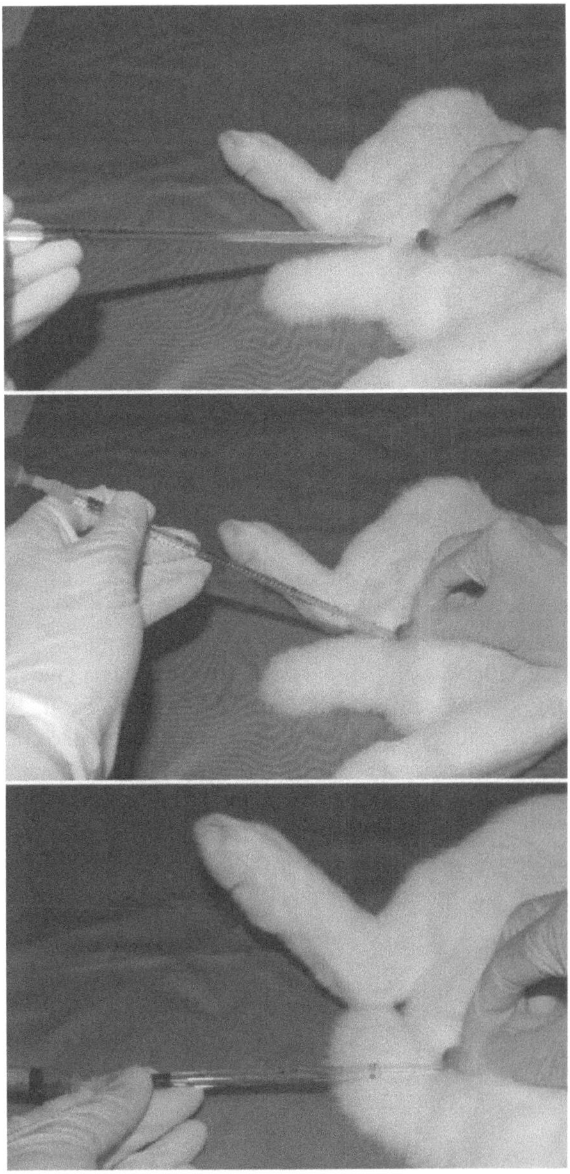

Fig. 2.3 Artificial insemination (AI). *Top*: In the photograph, rabbit is anesthetized by thiamylal sodium which is an ultra-short acting anesthetic. When insemination is done by two persons, one retains rabbit by holding her back and the other should operate the glass tube. *Middle*: The glass tube connected with 1 ml-syringe is inserted into the vagina. *Low*: The pipette should be inserted 10–15 cm into the vagina to ensure the delivery of sperm to female rabbits, because the vagina of rabbit is long and there is distance to orifice of uterus. At the time of insemination, ovulation of the rabbits must be induced by injection of 50 U of hCG

that a better performance is generally observed with AI than with natural mating. Judging from our result in JW rabbits, such as mean value of sperm volume, sperm concentration and motility, one ejaculate form one male rabbit can be sufficient to fertilize about 25 female rabbits using AI.

2.3 Induction of Parturition by Oxytocin

Induction of parturition by oxytocin can be helpful to decrease the accidental death of newborn pups at birth. Primiparous and poorly nursing rabbits often show poor results at birth. In these cases, it is better to inject oxytocin to induce parturition under our control. We usually inject subcutaneously 5U of oxytocin to the rabbit 30 days after AI.

2.4 Nursing of Suckling Pups

JW rabbits are very obedient and easy to be managed and they have shown good performance in reproduction efficiency in our facilities. But we can increase the weaning rate of newborn pups and rescue important ones if we make in due in time specific effort to care them rather than leaving rabbits alone nursing their progeny. In general, death of newborn pups occurs within 1 week. The most likely reasons of the death are thought to be on one hand an insufficient suckling and on the other hand crashing to death by the mother when it enters the nest box. Therefore, the status of suckling by the pups should be watched carefully for the first week after birth (Fig. 2.4). It is known that mother rabbits give milk to pups only one time per day, so they can be separated from the pups for the major part of the day. This is effective for preventing the accidental death due to crashing by parents. In this case, you can transfer nest box into the parental cage once per day and let mother rabbit give milk to suckling pups. If the lactation is not enough, you can inject oxytocin to mother rabbit. These methods for nursing of newborn pups may however induce neglect and cannibalism in the mother as adverse effects, especially in nervous rabbit strains, may take place due to the increase of human intervention.

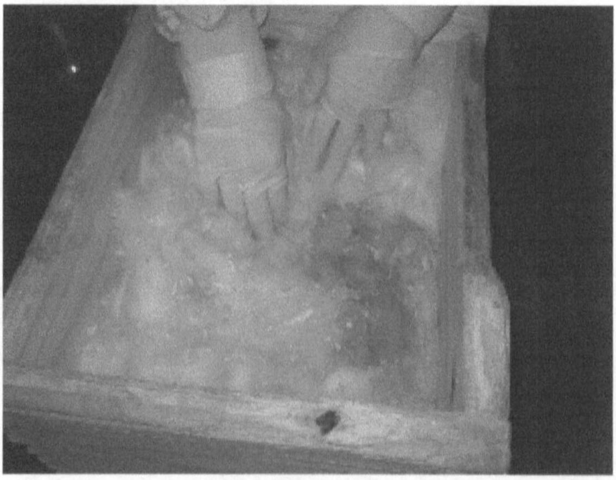

Fig. 2.4 Nest box and newborn pups. In our laboratory, to increase weaning rate of newborn pups, we usually watch the status of suckling by the pups for the first week after birth

2.5 Storage of Rabbit Semen

For temporally storage, if spermatozoa is kept in appropriate buffer and temperature, we can maintain it for several hours to a few days with normal fertility (Lopez-Gatius et al. 2005; Roca et al. 2000). Roca et al. (2000) reported the viability and fertility of rabbit sperm diluted in TCG buffer and stored at 15°C. They showed the TCG buffer is effective for retaining the fertilizing capacity of rabbit spermatozoa stored at 15°C up to 48 h (Roca et al. 2000). Moreover, Lopez-Gatius et al. (2005) reported that rabbit spermatozoa were effectively stored in the solid state by adding gelatin to the buffer, at 15°C with fertility preserved for up to 5 days. On the other hand, the method for the storage of rabbit semen is cryopreservation in liquid nitrogen. Cryopreserved spermatozoa can be maintained stably for a long time. This allows to decrease the number of living rabbits to reduce the cost and space in comparison with the maintenance of living individuals. Until now, several studies on cryoprotectants and protocols for cryopreservation of rabbit spermatozoa have been reported (Chen et al. 1989; Dalimata and Graham 1997; Fox and Burdick 1963; Kashiwazaki et al. 2006; O'shea and Wales 1969; Stranzinger et al. 1971).

2.6 Rabbit Semen Freezing

2.6.1 Cryoprotectants for Rabbit Semen

Until now, several cryoprotectants for rabbit sperm such as glycerol (Stranzinger et al. 1971), ethylene glycol (Fox and Burdick 1963), dimethylsulfoxide (DMSO) (Vicente and Viudes-De-Castro 1996) and acetamide (Chen et al. 1989; Dalimata and Graham 1997) have been reported and motile sperm having fertility was obtained after freeze-thawed using these cryoprotectants. Glycerol is the most frequently cryoprotectant used for domestic animals such as cattle, but it was reported that glycerol was not a suitable cryoprotectant for rabbits because it leads to a low fertility in AI after thawing (Curry 2000). Recently, Kashiwazaki et al. (2006) reported that lactamide and acetamide are suitable cryoprotectants for rabbit sperm comparing with glycerol and DMSO in JW rabbits. In our laboratory, we use egg-yolk acetamide (EYA) solution as a cryoprotective solution for rabbit sperm freezing, referring to the composition reported by Chen et al. (1989) and Dalimata and Graham (1997).

2.6.2 Protocol of Rabbit Semen Freezing Using Egg-Yolk Acetamide Extender

A protocol for the freezing of rabbit sperm using EYA extender in our laboratory is referred below. At first, semen is collected using AV, and the sperm concentration and motility is counted under microscope. The sperm concentration higher than 600 × 106/ml is adjusted to 600 × 10^6/ml with TCG buffer. Then, semen is diluted to six times in EYA extender and cooled with programmed cooling machine from room temperature to 5°C at a rate of –0.2°C/min for nearly 2h. After cooling, sperm is packed into straws in size of 0.5ml and it is kept more 30min in a refrigerator at 5°C. Freezing of sperm is performed by maintaining straws in the vapor of liquid nitrogen at 4–5cm above surface of liquid level for 15min, and straws are plunged into liquid nitrogen to finish it completely (Fig. 2.5). After freezing, the motility of sperm after thawing should be confirmed using at least one straw. In our laboratory, we keep sperm showing more than 30% in motility after thawing.

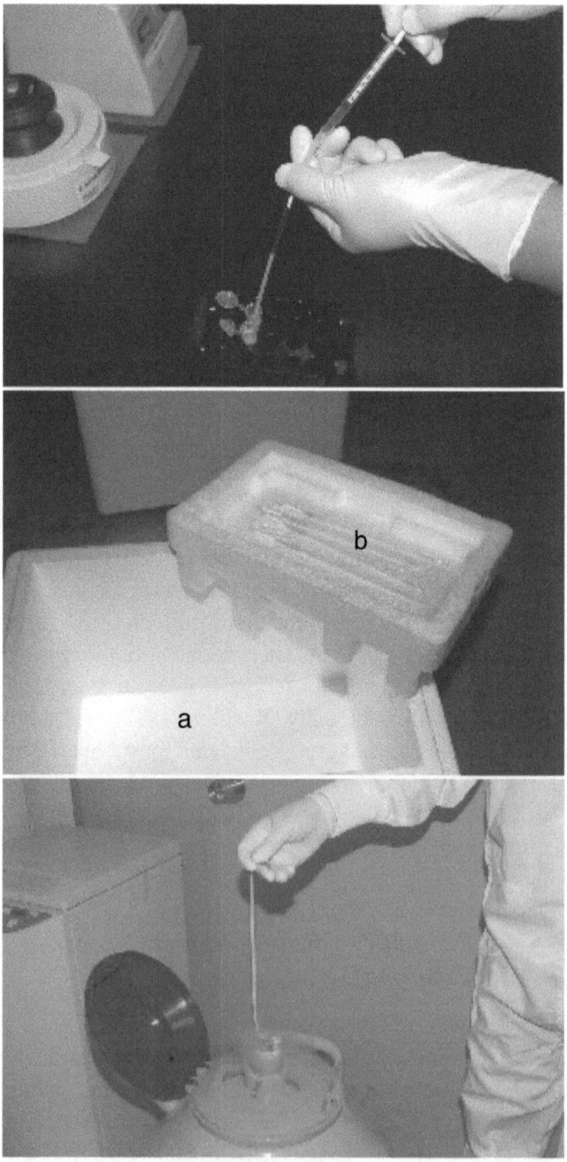

Fig. 2.5 Rabbit semen freezing. *Top*: After dilution of semen with EYH extender and cooling to 5°C, sperm is packed into straws in size of 0.5ml. *Middle*: For rabbit semen freezing, we use styrofoam box (**a**) and float (**b**). The sickness of the float is adjusted to appropriate size to keep at 4–5cm above surface of liquid level when it is put on liquid nitrogen. *Bottom*: After freezing, straws are kept in liquid nitrogen tank

2.6.3 AI Using Frozen and Thawed Rabbit Spermatozoa

For thawing, the frozen straws are dipped in water bath at 37°C for 30 s. The frozen-thawed sperm may be used just for AI. By the combination of cryopreservation and AI, O'Shea and Wales reported (1969) high fertility (74%) but slightly low litter size (4.6 pups). On the other hands, Viudes-de-Castro et al. (1996) reported similar performance of litter size in comparison with frozen and fresh semen. In our hands, in fact, when the same number of motile spermatozoa was used for AI with both frozen and fresh semen, the pregnancy rate was not significantly different but litter size with the frozen material was significantly lower than that in fresh semen. With our protocol, the efficiency of AI using frozen-thawed sperm with 20–40 × 10^6 spermatozoa/doe (n = 68), the pregnancy rate and mean litter size was 72.1% and 4.3 ± 2.8 pups, respectively. However, this technique of combination of semen freezing and AI can be sufficiently useful for preservation of rabbit strain such as important human disease model animal because of low cost and stability in the long term. Recently, its usefulness was demonstrated in a practical use such as re-establishment of transgenic rabbit lines (Blash et al. 2005) and transportation between the two counties beyond the ocean (Liu et al. 2007).

References

Blash, S., Chen, L., Harvey, M. and Gavin, W.G. (2005) Reestablishment of a transgenic rabbit line by artificial insemination using cryopreserved semen. Lab Anim 34:61–63.

Castellini, C. and Lattaioli, P. (1999) Effect of number of motile sperms inseminated on reproductive performance of rabbit does. Anim Reprod Sci 57:111–120.

Chen, Y., Li, J., Simkin, M.E., Yang, X. and Foote, R.H. (1989) Fertility of fresh and frozen rabbit semen inseminated at different times is indicative of male differences in capacitation time. Biol Reprod 41:848–853.

Curry, M.R. (2000) Cryopreservation of semen from domestic livestock. Rev Reprod 5:46–52.

Dalimata, A.M. and Graham, J.K. (1997) Cryopreservation of rabbit spermatozoa using acetamide in combination with trehalose and methyl cellulose. Theriogenology 49:831–841.

El-Gaafary, M.N. (1994) Quality and fertility of cooled rabbit semen supplemented with cyclic-AMP stimulators. Anim Reprod Sci 34:307–313.

Fox, R.R. and Burdick, J.F. (1963) Preservation of rabbit spermatozoa: ethylene glycol vs glycerol for frozen semen. Proc Soc Exp Biol Med 113:853–856.

Kashiwazaki, N., Okuda, Y., Seita, Y., Hisamatsu, S., Sonoki, S., Shino, M., Masaoka, T. and Inomata, T. (2006) Comparison of glycerol, lactamide, acetamide and dimethylsulfoxide as cryoprotectants of Japanese white rabbit spermatozoa. J Reprod Develop 52:511–516.

Liu, E., Kitajima, S., Wiese, E., Reifenberg, K., Morimoto, M., Watanabe, T. and Fan, J. (2007) Re-establishment of complement C6-deficient rabbit colony by cryopreserved sperm transported from abroad. Exp Anim 56:167–171.

Lopez-Gatius, F., Sances, G., Sancho, M., Yaniz, J., Santolaria, P., Gutierrez, R., Nunez, M., Nunez, J. and Soler, C. (2005) Effect of solid storage at 15 degrees C on the subsequent motility and fertility of rabbit semen. Theriogenology 64:252–260.

Morrell, J.M. (1995) Artificial insemination in rabbits. British Vet J 151:477–488.

Naughton, C.K., Nelson, D.R. and Thomas, A.J., Jr. (2003) Development of an inexpensive artificial vagina for semen collection from rabbits. J Androl 24:712–715.

O'shea, T. and Wales, R.G. (1969) Further studies of the deep freezing of rabbit spermatozoa in reconstituted skim milk powder. Aust J Biol Sci 22:709–719.
Roca, J., Martinez, S., Vazquez, J.M., Lucas, X., Parrilla, I. and Martinez, E.A. (2000) Viability and fertility of rabbit spermatozoa diluted in Tris-buffer extenders and stored at 15 degrees C. Anim Reprod Sci 64:103–112.
Stranzinger, G.F., Maurer, R.R. and Paufler, S.K. (1971) Fertility of frozen rabbit semen. J Reprod Fertil 24:111–113.
Vicente, J.S. and Viudes-De-Castro, M.P. (1996) A sucrose-DMSO extender for freezing rabbit semen. Reprod Nutr Dev 36:485–492.
Viudes-De-Castro, M.P. and Vicente, J.S. (1997) Effect of sperm count on the fertility and prolificity rates of meat rabbits. Anim Reprod Sci 46:313–319.
Viudes-De-Castro, M.P., Moce, E., Vicente, J.S. (1996) A simple method for freezing rabbit semen with successful results on fertility and prolificity. Anim Reprod Sci 44:195–201.

Chapter 3
Basic Methods for Experimental Rabbits

Kazutoshi Nishijima

Abstract To obtain reliable data from animal experiments, proper skills must be acquired. Experimental methods are different in accordance with animal species to treat. Restraining, compound administrating, sampling and anesthetizing methods specialized for experiment rabbits are described in this chapter.

Keywords Restraint, Compound administration, blood sampling, anesthesia

3.1 Introduction

Reliable and reproducible experimental results must be followed by proper experimental methods, which will additionally lower distress of the rabbits. To perform the proper experimental procedures, anatomical, physiological and behavioral characteristics of the rabbits are necessary to be learned. Some trainings with skills handling rabbits are needed.

3.2 Handling and Restraint

Rabbit is generally gentle animal and relatively easy to handle. However, it should be careful when handling rabbits during lactation period, since nursing mothers can be aggressive.

K. Nishijima (✉)
Division of Biological Resources and Development, Analysis Research Center for Experimental Sciences, Saga University, Saga, Japan
e-mail: Kanish@cc.saga-u.ac.jp

L.-M. Houdebine and J. Fan (eds.), *Rabbit Biotechnology: Rabbit Genomics, Transgenesis, Cloning and Models*,

3.2.1 Physical Methods

Rabbits should be handled gently but firmly. A rabbit should be picked up with a wide grab on the loose skin over its back at posterior of the shoulder with one hand (Fig. 3.1a) for transportation. The rabbit can be more stabilized when hind its hip is held with the other hand (Fig. 3.1b). Holding legs is not preferable because the rabbit try to kick and become unruly. The hip of the rabbit can be stuck between thighs of handler who is sitting on a chair, and forefeet are hold upward by his hand(s) (Fig. 3.2).

These methods to restrain the rabbit need the handler's skill, and not adequate to restrain the rabbit for a long time period.

3.2.2 Mechanical Methods

Several types of rabbit-restraining devices are commercially available. Rabbit must be handled gently not to be excited when they are put into the restraining devices. Care must be taken to prevent injuring the rabbit by impacts or nipping with hard materials like metal frames of the restraining devices. The rabbit-restraining devices, formed box, tubular (Fig. 3.3) or board (Fig. 3.4) should be selected by the methodology of the experiment.

These methods are well adapted to restrain the rabbit convincingly for a long time. However, since it takes time to put the rabbits into the restraining devices, these methods are not adequate when you need to handle numbers of rabbits just for a brief procedure like single administration of compound or blood sampling.

Swathing the rabbit in a cloth is another option to restrain (Fig. 3.5). The rabbit is hardly move wrapped with the cloth, however, it is needed to hold the cloth swathing the rabbit sometimes. Needed part of body, like the ears, is protruded out through intervals of the cloth.

a b

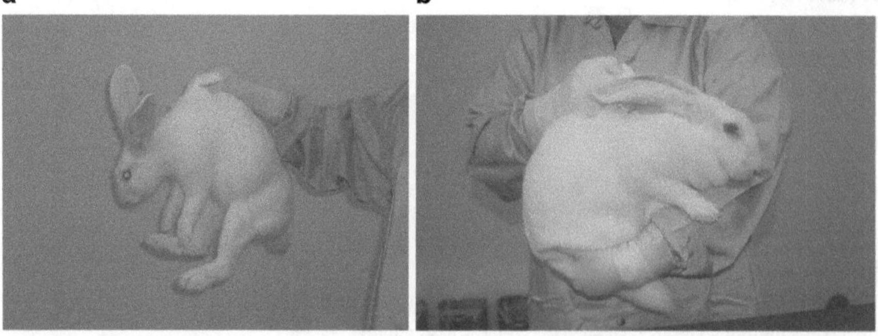

Fig. 3.1 Hand restraint for transporting

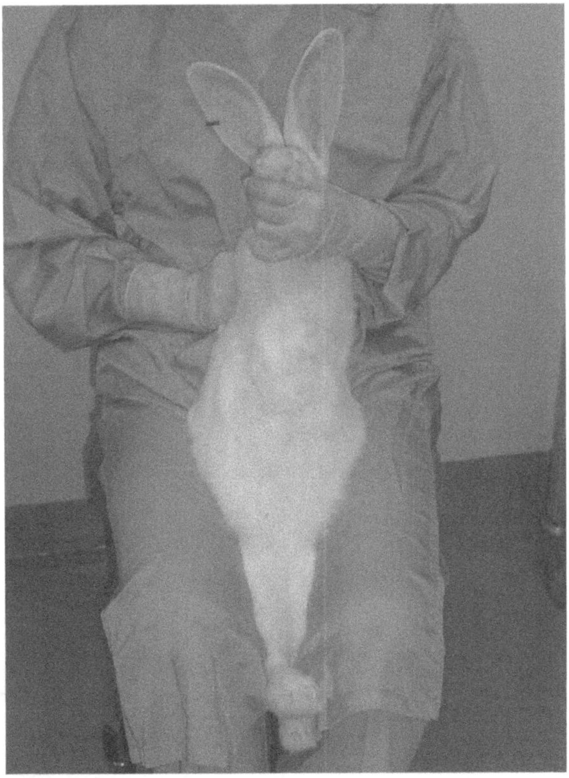

Fig. 3.2 Physical restraint

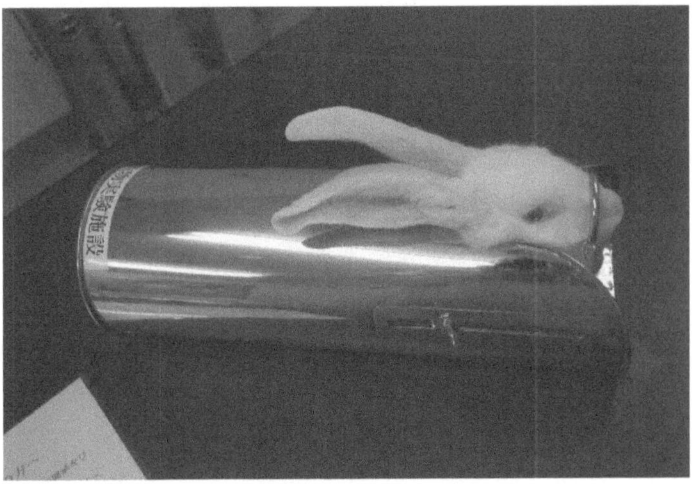

Fig. 3.3 Restraint tube

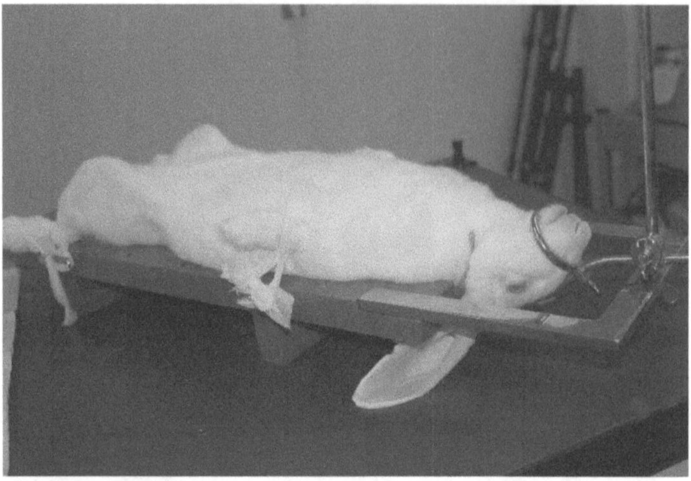

Fig. 3.4 Restraint board

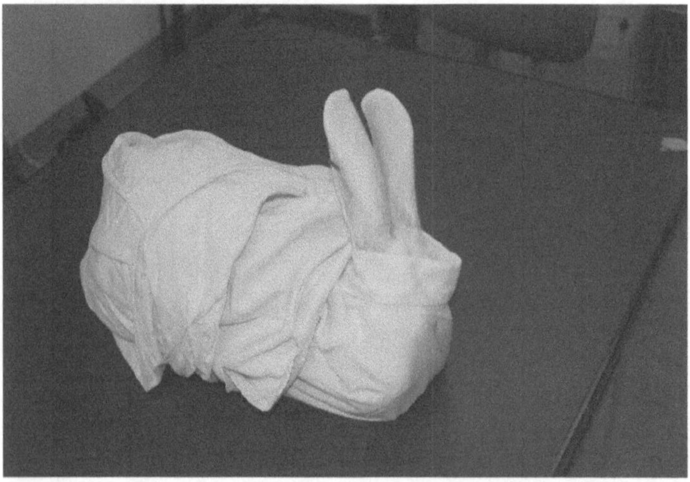

Fig. 3.5 Restraint with a cloth

3.3 Compound Administration

3.3.1 Oral

Placing a compound directly into the stomach with a catheter is the most accurate method for oral compound administration. It is necessary to restrain the rabbit for the oral compound administration. The forefeet and roots of the ears of the rabbit should be hold tightly and put upward, the hip of the rabbit can be stuck between thighs of handler. A rabbit-restraining device for oral compound administration is also available. Mouth movement is controlled with a mouth speculum, and the tongue of the rabbit is placed under the speculum.

3.3.2 *Intravenous*

The helicine branch of caudal auricular vein of the rabbit's ear is choice of intravenous injections, since it is large enough, linear course and readily accessible. The ear should be wiped with ethanol-absorbed cotton for cleaning, sterilization and venous dilatation. Picking coat around the vein makes the vein well dilated and easily viewable. Air bubbles must be cleared away from inside of the syringe to prevent air emboli in the rabbit's vein. The needle is inserted in the vein in the direction of the ear root (afferent, Fig. 3.6). Before injection, assure that the needle was put in the vein, by confirming the blood flows in the syringe when the syringe plunger is withdrawn. After injection, apply firm pressure on the needle insertion site to treat bleeding.

3.3.3 *Intramuscular*

The injection site should be a body of a large muscle, like gluteal or thigh muscles. The needle is inserted vertically to the skin. Confirm that no blood can be aspirated when syringe plunger is withdrawn lightly before injection. The injected muscle should be massaged to diffuse the reagent.

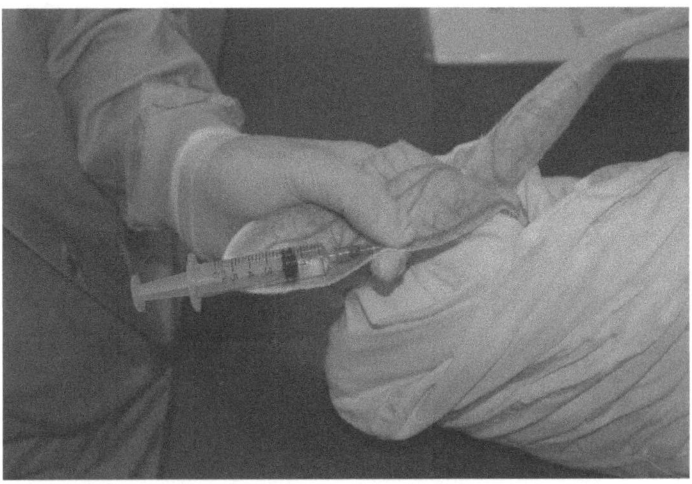

Fig. 3.6 Compound administration via a vein

3.3.4 Intraperitoneal

The rabbit should be restrained in posture with its hindquarter elevated to locate the intra-abdominal organs in near thoracic cavity, and prevent the organs injured with the needle. Insert the needle just lateral to the midline at the angle of approximately 30°, and then bring the needle forward subcutaneously. Thrust the needle at the angle of 45° cautiously to penetrate the abdominal wall, and inject the regent in the abdominal cavity.

3.3.5 Subcutaneous

Back of neck, lumbar portion of back, abdominal and inguinal regions are options of the site of the injection. Pickup the skin and insert the needle parallel to the body axis (Fig. 3.7). Confirm that the tip of the needle does not protrude outside the skin when it is swung, and inject the reagent. The injection site should be pinched to avoid outflow of the injected reagent.

3.3.6 Intradermal

The injection sites are commonly located in regions of back or flank that the rabbit cannot easily reach with his hind feet. The injection area should be clipped carefully, and a depilatory cream is applicable to remove hair. Tighten the skin of the

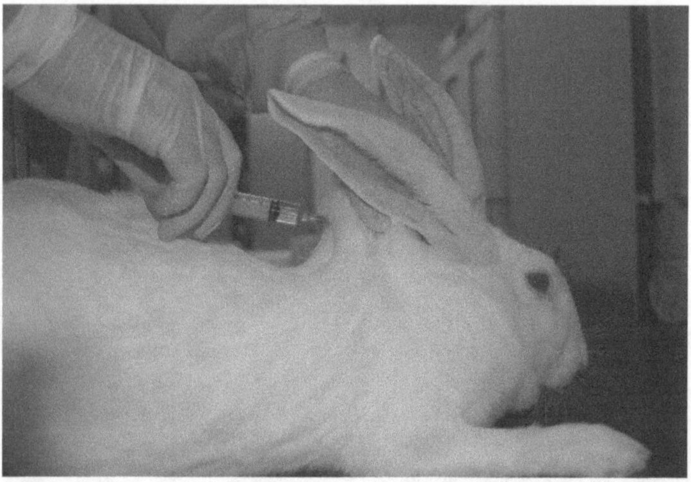

Fig. 3.7 Subcutaneous compound administration

injection site, and insert the needle parallel to the skin. When injected into dermis, a small bleb is generated. To avoid the outflow of the injected reagent, hold the needle for a few second.

3.4 Sampling Techniques

3.4.1 Blood Sampling

Blood vessels of the ears are commonly used for usual blood sampling, since they are readily accessible and large enough. Both of veins and arteries are available, and the helicine branch of caudal auricular veins and intermedial branch of caudal auricular arteries can be the first choice for the blood sampling. However, it is hard to collect large amount of blood within the veins because they collapse easily. Larger amount of blood can be collected from the arteries. Restrain the rabbit properly, and its ear should be wiped with ethanol for sterilization and dilatation. The dilatation can be induced by tapping or rubbing the ear or the site of the vessel. 22- to 25-gauge hypodermic needle is inserted into the vessel efferently in the vein or afferently in the artery (Fig. 3.8). Puncture needles with a catheter for chronic cannulation are more convenient for larger volume or repeating blood sampling. In such cases, attention must be paid if the catheter gets clogged with clotted blood. Large (20-gauge) catheter should be chosen, and maintain the blood flow through it. When the catheter is clogged, clean up the clot with the needle or something thready. After removing the catheter, apply firm pressure on the needle puncture site with dry cotton.

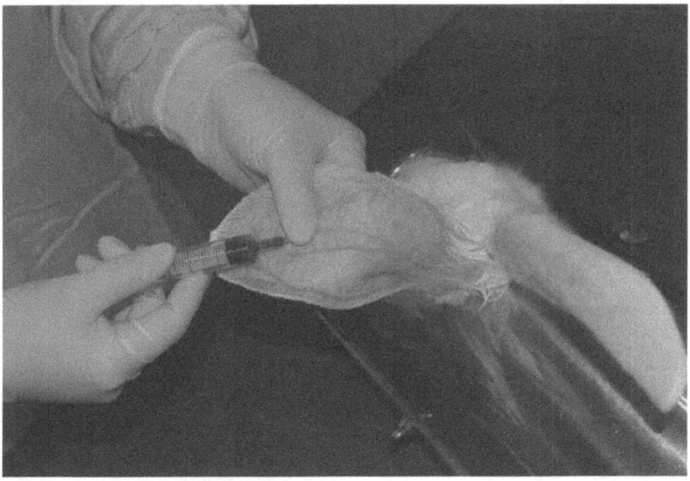

Fig. 3.8 Blood sampling via an intermedial auricular artery

Whole blood collection and exsanguination are sometimes done by cardiac puncture or carotid artery cannulation. Lay deeply anesthetized rabbit on its back upon a rabbit-restraining board. In the cardiac puncture, confirm the position of the heart by feeling the heart beat with a hand, and insert the needle through left intercostal space or from posterior of the xiphoid process (Fig. 3.9). Once the needle penetrates the heart, the blood flows into the syringe. In carotid artery cannulation, cervical region of the rabbit should be shaved and sterilized. The skin of the cervical region is incised along the midline, and the carotid artery running parallel to a trachea is exposed. Isolate the artery form surrounding connective tissue without incisive instruments not to injure nerves or small arteries. The peripheral side of the artery is tightly ligated with a suture thread, and the central side is pinched with a clamp to stop blood flow. Make an incision on the arterial wall between the ligated thread and the clamp. Inset an appropriate catheter, in diameter and length, into the artery toward the heart. The catheter should be fixed by a suture thread ligated over the arterial wall, then remove the clamp.

3.4.2 Urine Collection

There are two methods for urine collection. One is forced urine collection which is using insertion of a catheter through the urethra. A male rabbit should be hold in a sitting position with clutching the rabbit's front and hind legs of each side with each hand. Expose the penis, and lubricated flexible catheter should be inserted carefully into opening of the urethra (Fig. 3.10). When the tip of the catheter gets to the bladder after passing a few centimeters through the urethra, urine will flow.

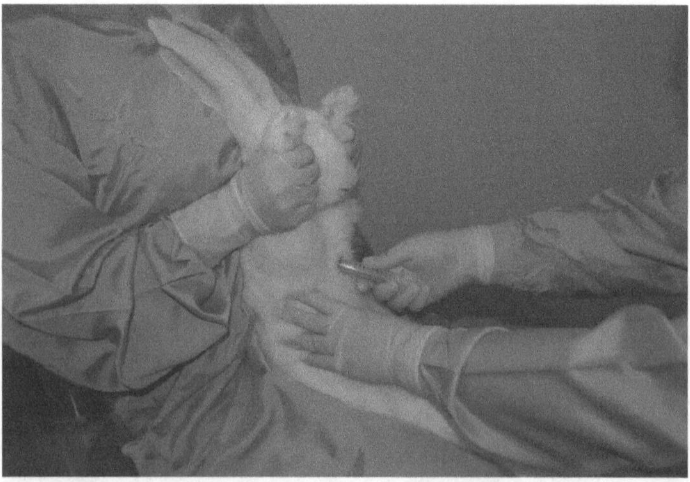

Fig. 3.9 Blood sampling via a heart

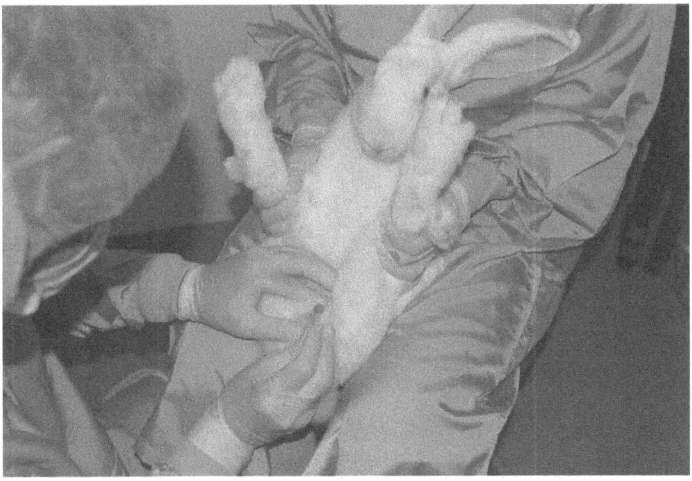

Fig. 3.10 Urine collection

The other option is naturally collection method using a metabolic cage. This method should be better for female rabbit or the urine collection over long time period. However, some considerations should be necessary on urine parameter, because contaminations or alterations of components in the urine might occur before measuring.

3.5 Anesthesiology

Though the anesthesia is indispensable for experiments using the rabbit, a particular attention must be paid for application of it. The rabbit is susceptive to stresses, and sometimes this leads side-effects of the anesthetic agents. Administration of single anesthetic by a high dosage raises incidence of anesthetizing accidents by the side-effects. Thus, an application of a preanesthetic agent like xylazine (5 mg/kg) or medetomidine (0.1–0.5 mg/kg) and combined usage of multi agents to reduce the dosage of each ones are recommended. There are numbers of combinations of anesthetic agents which are suggested being applicable for rabbit, and some of them are listed in Table 3.1. Selection of the anesthetic should be done by the experimental procedure.

Since some anesthetic agents cause the rabbit a breathing trouble during its induction, such agents must be administrated slowly with special care. The depth of anesthesia is determined by pinching of the ear or flexion reflex of the leg. Airway should be controlled with a tracheal cannulation or an artificial respiratory mask for long term anesthesia. It is important to control the body temperature of the rabbit because it has relatively high average temperature (38.5–40.0°C).

Table 3.1 Typical anesthetic drugs used for rabbits (Extracted and modified from Manning et al. 1994)

Drug	Dosage	Route	Duration
Ketamine	20–50 mg/kg	IM/SC	30 min
+ xylazine	5–10 mg/kg		
Ketamine	25–35 mg/kg	IM/SC	90–180 min
+ medetomidine	~0.5 mg/kg		
Ketamine	15–30	IM/SC	20 min
+ midazolam	2–5		
Thiamylal sodium	12–25 mg/kg	IV	15 min
Isoflurane	3–5% (induction) 1–3.5 (maintenance)	Inhalation	

3.6 Euthanasia

When the rabbits are put to death for rearing management or experimental design, the euthanasia must be accomplished appropriately. The procedure of the euthanasia consists mainly of rapid death with a minimum fear and pain for the rabbit.

The most common procedure is excess administration of anesthetic agents. Intravenous or intraperitoneal (for young small rabbits) administration is a usual route for the anesthetic agents for euthanasia. Inhalation of carbon dioxide is also preferred for the euthanasia. The rabbit is put in an airtight space which is subsequently filled with carbon dioxide. Death of the rabbit must be confirmed by stopping of the heart beat.

The euthanasia should be accomplished in a shade of obscurity, considering emotions of other people.

Reference

Manning PJ, Ringler DH, Newcomer CE (eds): The Biology of the Laboratory Rabbit (ed 2). Academic, San Diego, CA, 1994

Chapter 4
Useful Information for Rabbit Genes, Proteins, and Antibodies

Tomonari Koike, Jifeng Zhang, and Jianglin Fan

Abstract In this chapter, we have collected useful website information on rabbit research. Readers can access and refer to this information on rabbit gene and protein sequences along with primary antibodies against rabbit.

Keywords Website, rabbit genome, antibodies, physiology

4.1 Introduction

For many experiments with rabbits, genetic information and research tools such as antibodies and ELISA kits are essential for analyzing gene and protein expression and performing immunohistochemical staining. For gene-targeting in rabbits, genome information is critical. In this chapter, we will introduce such information, many of which can be accessed by website searching. We will include the database of the rabbit gene and proteins, commercially available antibodies against rabbit proteins and ELISA kits for measurement of different rabbit plasma proteins. Finally, we introduce the websites for the physiological information of the rabbit so that researchers can compare their results derived from rabbits with humans. Comprehensive information on the rabbit genome has also been described in other chapter elsewhere.

4.2 Rabbit Genome, cDNA, and Protein Sequences

National Center for Biotechnology Information provides almost all information on rabbit genome (http://www.ncbi.nlm.nih.gov/projects/genome/guide/rabbit/). You can also find rabbit genome sequences from the Broad Institute, the Lund University

T. Koike (✉)
Department of Molecular Pathology, Interdisciplinary Graduate School of Medicine and Engineering, University of Yamanashi, Yamanashi, Japan
e-mail: koiket@yamanashi.ac.jp

L.-M. Houdebine and J. Fan (eds.), *Rabbit Biotechnology: Rabbit Genomics, Transgenesis, Cloning and Models*,

and Ensemble genome browser by Institut national de la researche agronomique. Furthermore, you can browse the rabbit scaffolds at UCSC Genome Bioinformatics website (http://genome.ucsc.edu). Furthermore, you can find the ENCODE Project, BAC library from the BACPAC Resource Center (BPRC, http://bacpac.chori.org), rabbit mapping resources (Rabbitmap, http://locus.jouy.inra.fr/cgi-bin/lgbc/mapping/common/intro2.pl?BASE = rabbit). The most important feature of this website is that you can link to many rabbit community resources such as NIAID rabbit immunology and infectious disease research, ImMunoGenetics information system, American Rabbit, Breeders Association, Recognized rabbit breeds, FAO Report on Rabbit Husbandry, Health, and Production, Online Mendelian Inheritance in Animals and many diverse information on rabbits on Animal Diversity Web.

Rabbit information (nucleotides, proteins, structure, genome sequences and so on) has been assembled together in the Taxonomy browser – rabbit (Oryctolagus cuniculus) (Table 4.1), but so far (until 2008, September), there is only 1,179 genes in Database name-Gene. There are about 34,918 EST, 7,630 proteins sequences exist in NCBI database. In the GSS division of GenBank in NCBI, you can find 1,305 rabbit genome sequences, which is from Rabbit BAC Library. In addition, you can find 334 rabbit protein structure information from NCBI database. The rabbit cDNA libraries are commercially available from Stratagene company (www.stratagene.com). There are seven different libraries shown in their website, including heart, liver, kidney, muscle, etc.

4.3 Primary Antibodies Against Rabbit Proteins

The rabbit is one of the most used animals for the generation of both polyclonal and monoclonal antibodies against many species proteins; however, the primary antibodies specifically against *rabbit* proteins are still limited. To overcome this problem, we always use the antibodies against human proteins (e.g. α- and β-actin) to that can cross-react with rabbit proteins because many proteins between two species are highly conserved. There are several companies that manufacture and sell the primary antibodies against rabbit for flow cytometry, Western blots, and immunohistochemical staining. The most popular antibodies against rabbit are antibodies against rabbit macrophages: RAM11 (by rabbit alveolar macrophage extracts, Dako) and RBM2 (by thioglycollate-elicited macrophages, Cosmos Bio) because rabbits are frequently used as a model for the study of atherosclerosis. Spring Valley Laboratories, Inc. provides a panel of monoclonal antibodies against rabbit proteins that include immunoglobulin isotypes and light chains (IgG, IgA, IgM, Kappa light chain), histocompatibility antigens (MHC class I, MHC class II, MHC class II DR, MHC class II DQ), integrins and cellular adhesion molecules (CD11a, CD11b, CD11c, CD18, CD43, CD58), cellular differentiation markers (CD3, CD4, CD5, CD8, CD9, CD25, CD44, CD45, CD79a) and lymphocyte proteins (Thy-1, GALT lymphocytes, and T lymphocytes). In addition, a panel of monoclonal antibodies against rabbit lymphocyte antigens including T cells, B cells, macrophages,

Table 4.1 Useful websites for rabbit research

Website name	Address	Description
GENETIC DATABASE		
1. NCBI, Taxonomy browser, Oryctolagus cuniculus	http://www.ncbi.nlm.nih.gov/Taxonomy/Browser/wwwtax.cgi?mode=Info&id=9986&lvl=3&lin=f&keep=1&srchmode=1&unlock	You can find the table, named 'Entrez records' which shows the number of all nucreotides, proteins and genome sequences of rabbit in NCBI. In 'External Information Resources (NCBI LinkOut)', it links to the famous sites about rabbit taxnomy and phylogenetic information.
2. Rabbitmap database (INSTITUT NATIONAL DE LA RECHERCHE AGRONOMIQUE)	http://locus.jouy.inra.fr/cgibin/lgbc/mapping/common/intro2.pl?BASE=rabbit	This is the database site of loci and sequences of rabbit genes. You can find partial genome sequences in some genes.
3. NCBI, Rabbit genome resources	http://www.ncbi.nlm.nih.gov/projects/genome/guide/rabbit/	Newest informations of genes, proteins and papars about rabbits are updated in 'New This Month In'. In 'Sequence Resources', it links to the sites of the institutes of genome project. 'Community resources' shows the sites of rabbit immunology, breeding and inheritances.
ANTIBODIES / ELISA KITS		
4. Spring Valley laboratories, Inc.	http://www.svlab.com/antibody.php	They produce and sell the mouse-monoclonal antibodies anti-rabbits' immunological proteins, e.g., immunoglobulins. MHCs, CDs.
5. Research diagnostics, Inc.	http://www.researchd.com/rbcds/rbcdmono.htm	They sell the mouse-monoclonal antibodies anti-lymphocyte antigens of rabbits (CD4, CD8, B cell, etc.).
6. DaLiang Fanbang	http://zdsj.com/rabbit.htm	There is a list of >80 kinds of ELISA kits for rabbit proteins.
7. Biocompare Buyer's Guide	http://www.biocompare.com/jump/2045/Antibodies.html	This is a useful web site for searching antibodies. You can select rabbit in 'Reactivity' and find antibodies for rabbit.
PHYSIOLOGY		
8. TestDiet®	http://www.testdiet.com/	TestDiet® is the company selling diets for laboratory animals. You can find the basic data of rabbit physiology referred to the "Guide for the Care and use of Laboratory Animals" – NIH Publication No. 85-23, Revised 1985.

neutrophils, CD4, CD5, CD8, CD9, CD11a, CD11B, CD11C, CD18, CD25, CD28, CD43, CD44, CD51/61. CD58 can be obtained from Research Diagnostics and Acris Antibodies GmbH.

Rabbit is a good model for inflammation research, so rabbit CRP and active PAI-1 ELISA kits are produced by several companies. Dalian Fanbang Company in China sells more than 80 rabbit ELISA kits for measuring inflammatory cytokines, growth factors, MMPs, and soluble adhesion molecules. Biocompare Buyer's Guide is a useful website by which you can find any commercially available antibodies you want. When you click 'rabbit' from 'the reactivity', you can find first anti-rabbit antibodies and/or any antibodies cross-reacted with rabbit proteins.

4.4 Other Useful Websites for Rabbit Physiology

Information about "Care and Feeding & Physiology" and special diets for experiments (such as high fat diets and cholesterol-rich diets for producing diabetes and atherosclerosis) can be found in Test Diet (http://www.testdiet.com/). Information on rabbit health, diseases, care, feeding and shopping can be found in Animal Hospitals (http://www.animalhospitals-usa.com/small_pets/rabbits.html) and common diseases of rabbits have been introduced (https://secure.lifelearn.co.uk/store/product_samples/pdfs/ch007.pdf).

Chapter 5
General Physiology of Rabbits

Masashi Morimoto

Abstract In this chapter, I summarize the basic and general physiology of experimental rabbits. The information should be helpful for researchers who use rabbits as an experimental tool. The emphasis will be given to the comparison between rabbits and humans.

Keywords Comparative medicine, normal rabbit values, physiology

5.1 Delivering and Nursing Physiology (Tables 5.1–5.3)

Rabbits do not show regular estrous cycles (Altman and Dittmer 1964, Harkness and Wagner 1983). Therefore, the ovulation can be induced either by copulation or by hormone injection. Fertilization will take place after 8–10 h. Gestation is about 28–36 days (Batchelor 1999; Bell 1999). The average litter size is about 4–10. The pups need to wean at about 1 month. If the pups are removed immediately after delivery, the mother rabbit will be sexually receptive at about a month. The new-born pups have little hair and they are deaf and blind. The fat content of the neonatal rabbits is much higher than that of the neonatal rats. The body weight at birth is around 65 g. After delivery, the mother rabbit returns to the nest once a day and nurses her pups. The nursing took place in the early morning and lasted for 2.7–4.5 min (Zarrow et al. 1965). The young rabbits start to take a solid food at about 20 days after birth. They are very mobile at 4 weeks of age and leave their nest. At that time (35 ~ 40 days after delivery), they start weaning, but they continue to suckle for some more weeks. The mother rabbits have maximum milk production 2 weeks after delivery. After weaning, lactation continues for an additional 2–4 weeks. At 8 weeks of age the young consume a solid food. For laboratory rabbits, weaning and separation from the mother can be achieved at 6–7 weeks of age (Hagen 1974).

M. Morimoto (✉)
Department of Rehabilitation, Kumamoto Health Science University, Kumamoto, Japan

L.-M. Houdebine and J. Fan (eds.), *Rabbit Biotechnology: Rabbit Genomics, Transgenesis, Cloning and Models*,

Table 5.1 Body weight of NZW

Sex/age	Weigth (kg)
♂♀0day	0.065
♂♀28 day	0.584
♂♀42 day	1.25
♂8 week	1.95 (1.60~2.30)
♀8 week	2.04 (1.50~2.50)
♂12 week	2.67 (2.30~3.00)
♀12 week	2.72 (2.10~3.00)
♂16 week	3.13 (2.60~3.50)
♀16 week	3.26 (2.60~3.70)
♂20 week	3.45 (2.80~3.90)
♀20 week	3.70 (3.00~4.30)
♂24 week	3.61 (3.00~4.30)
♀24 week	4.00 (3.40~4.80)

(Altman and Dittmer 1964)

Table 5.2 Biologic parameter of rabbits

Life span	5~7 years (up to 15 years)
Body Weight	2~5 kg (NZW)
Temperature	38°C~40°C
Mammary gland	8 or 10
Gestation	28~36 days
Litter size	4~10
Body weight at birth	30~100 g

Table 5.3 Milk composition

Water (%)	70~73
Protein (%)	10~15
Fat (%)	10~16
Carbohydrate (lactose) (%)	1.8~2
Ash (%)	2~2.5

(Harkness and Wagner 1983)

The eyes open at 10–11 days (Kersten et al. 1989). Hearing may develop at the same time. In laboratory rabbits, sexual maturity occurs at the age of 5–7 months in male and 9–12 months in female (Myers et al. 1994). Suckow et al. (2002) has described that smaller breeds typically reach puberty earlier, and larger breeds a bit later.

The normal rectal temperature of adult rabbit is 38–40°C. Rabbits can adapt to different ambient temperatures, but are susceptible to high temperature. When transportation of rabbits is made, we have to be careful with temperature. The ears represent a relatively large percentage of the body surface area (12%). The ears are important heat-sensing organs in the rabbit (Gonzalez et al. 1974). The metabolic rate varies widely between different species, being roughly related to the surface area of the mammal. The fasting metabolic rate of rabbit is 44 ~ 55 kcal/kg/day, or 26 ~ 38 kcal/m^2/h (Gillett 1994).

Water should always be available to rabbits. The daily water consumed by each rabbit is 200 ~ 500 ml. Rabbits consume approximately pellets by 50g/kg body weight. More water is needed for growing animals as well as pregnant and lactating females (Meredith 2000).

5.2 Behavioral and Sensory Physiology

It may be considered that the sight of a rabbit is not good. In visual field of a rabbit, horizontal one is very good, but vertical one is not. They have 180° field of vision, but cannot see area beneath their mouth. A range of stereoscopic vision is very small, so rabbits crash to an obstacle in their front. A rabbit is a nocturnal animal. A sight in a dark place is good, but it is generally considered that color sense is very bad. Hearing and sense of smell of a rabbit are very good. They use their whiskers and sensitivity of the lips as well as smell and taste during foraging (Meredith 2000). The whiskers are also used during orientation in the nests. Smell and taste is more important in identifying members of the own breeding group than vision. They have good hearing by reason of their big ears.

Rabbits have three specialized scent glands, i.e. in the anal region, in the groin and under the chin (Mykytowycz 1968). The territory of rabbits is marked by placing feces. The rabbit also marks its territory by pressing the under-chin against structures in its environment. The more dominant animals of a group have the larger anal glands. Males sprayed urine on females and young rabbits of their group for marking. Females had marking their young. They attack other young within the same group. Females may attack even their own young if they have been smeared with foreign urine (Mykytowycz 1968).

5.3 Urinary Physiology (Table 5.4)

The urine of adult rabbits is typically cloudy due to a relatively high concentration of ammonium magnesium phosphate and calcium carbonate monohydrate precipitates (Flatt and Carpenter 1971; Kaplan and Timmons 1979; Mitruka and Rawnsley 1977)). Rabbits fed a 10% calcium carbonate diet excrete about 60% of the calcium carbonate in their urine (Cheeke and Amber 1973). The urinary calcium level generally varies as the serum calcium level. A high calcium diet causes the urine to have a thick creamy appearance; however, the urine is viscous and pale even on diets not high in calcium (Hennemann 1959). The urine may also take on hues ranging from yellow or reddish to brown. In contrast, the urine of young rabbits is typically clear, although healthy young rabbits may have albuminuria. The urine is normally yellow but can also take on reddish or brown hues once they begin to eat green feed and cereal grains. Normal rabbits have few cells, bacteria, or casts in their urine (Suckow et al., 2002). The pH of the urine is typically alkaline at 7.6 ~ 8.8 (Gillett 1994). A normal adult rabbit produces approximately 20 ~ 350 ml of urine daily. Female rabbits have to urinate more copiously than male rabbits.

Table 5.4 Values of urine biochemical and enzyme parameters of NZW

Volume (ml/day)	20.0~350
Gravity	1.003~1.036
pH	7.60~8.80
Total protein (ml/day)	0.74~1.86
Urea nitrogen (g/day)	1.20~1.50
Uric acid (mg/day)	4.00~6.00
Creatinin (mg/day)	20.0~80.0
Ca (mg/day)	12.1~19.0
Cl (mg/day)	190~300
Mg (mg/day)	0.65~4.20
P (mg/day)	10.0~60.0
K (mg/day)	40.0~55.0
Na (mg/day)	50.0~70.0
Glomerular filtration rate (ml/min/kg)	7

5.4 Digestive Physiology (Table 5.5)

The rabbits chew their food with jaw movement up to 120 times/min. Because the cardiac sphincter is well developed in rabbits, rabbits cannot vomit (Gillett 1994; Harkness and Wagner 1983)).

The colon is divided into two portions (proximal and distal) by the fusus coli. Hard pellets comprise about two thirds of the fecal output. Cecotrophs (soft feces) was formed in the proximal colon and cecum. Cecotrophs are rich in nitrogen-containing compounds and vitamins, minerals as well as nitrogen. They are coated in a tough gelatin. Rabbits eat cecotrophs directly from the anus to obtain significant nourishment. Soft pellets are sometimes termed "night feces" (Suckow et al. 2002).

5.5 Respiratory Physiology

Rabbits have a well-developed sense of smell. Nasal breathing in rabbits is characterized by twitching of the nostrils at rates varying from 20 to 120 times per minute ((Table 5.6) (Harkness and Wagner 1983; Kaplan and Timmons 1979)). The twitching may be absent in the relaxed rabbit.

The pharynx of the rabbit is long and narrow. A position of vocal fold is deep. The tongue is relatively large. These features make endotracheal intubation difficult to perform in the rabbit.

Flow volume of air to the left lung is higher than to the right due to the lower resistance of the proximal airways per unit volume (Yokoyama 1979). In rabbits, lung volume increases with age in contrast to that of humans in which it decreases.

Because digestive organs are very large, a thorax cavity is very small. Therefore there is very little lung capacity of a rabbit and size of a heart is small in the ratio with the weight.

Table 5.5 Digestional parameter of rabbits

Fecal production (g/day)	15~60
Food consumption (g/day)	100~200
Water consumption (ml/day)	200~500
Gastrointestinal transit time (h)	4~5
Metabolic rate	26~38 cal/m^2/h 44~55 kcal/kg/day

Table 5.6 Circulatory and respiratory parameter of rabbits

Heart rate	200–300 beats/min
Respiratory rate	32–60 breaths/min
Arterial blood pressure	90–130 mmHg systolic 80–90 mmHg diastolic
Arterial blood pH	7.2~7.5
pO_2	85~102 mmHg
pCO_2	20~46 Torr
HCO_3	12~24 mmol/l
Arterial oxygen	12.6%~15.8% volume
Total lung capacity	111 ± 14.7 ml
Minute volume	0.6 l/min
Tidal volume	4–6 ml/kg

5.6 Cardiac Physiology (Table 5.6)

In rabbits, electric potential of electrocardiogram is low. During surgery operation, we observed arrhythmia, down of blood and pulse pressure in rabbits (Harkness and Wagner 1983; Kaplan and Timmons 1979)). The rabbit heart is relatively resistant to oxidative damage (Matsuki et al. 1990). The carotid artery of the rabbit is more compliant than that of the dog and has a greater ratio of elastin to collagen (Cox 1978). The blood pressure in the central ear artery of the Japanese white rabbit (about 3.0 kg) after neurectomy is 124 mm/Hg (systolic) and 83 mm/Hg (diastolic). The blood pressure increases with the growth of the body weight. Xylazine and ketamine modulate blood pressure as well as heart and respiratory rate (Sanford and Colby 1980).

5.7 Hematologic Values

Values for hematologic parameters are shown in Table 5.7 (Mitruka and Rawnsley 1977).

5.8 Serum Biochemical and Enzyme Values

Values of serum biochemical and enzyme parameters are shown in Table 5.8 (Mitruka and Rawnsley 1977; Wolford et al. 1986). These values can be used for reference since each colony may have different values dependent upon age, sex, and raising conditions. Each laboratory should establish its own normal values.

Table 5.7 Hematologic values of NZW

		♂	♀
RBC	$\times 10^6/\mu l$	6.70 ± 0.62	6.31 ± 0.60
Hb	g/dl	13.9 ± 1.75	12.8 ± 1.50
Ht	%	41.5 ± 4025	39.8 ± 4.40
MCV	fl	62.5 ± 2.00	63.1 ± 1.92
MCH	pg	20.7 ± 1.00	20.3 ± 1.60
MCHC	%	33.5 ± 1.85	32.2 ± 1.74
Blood volume	ml/kg	55.6	
Plasma volume	ml/kg	38.8	
RBC volume	ml/kg	16.8	
Reticulocyte (4–6 months)	%	3.6 ± 1.8	
WBC	$\times 10^2/\mu l$	90.0 ± 17.5	79.0 ± 13.5
Neutrophils	%	46.0 ± 4.00	43.4 ± 3.50
Eosinophils	%	2.00 ± 0.75	2.00 ± 0.60
Basophils	%	5.00 ± 1.25	4.30 ± 0.95
Monocytes	%	8.00 ± 2.00	9.00 ± 2.00
Lymphocytes	%	39.0 ± 5.50	41.8 ± 5.18
Platelets (4–6 mon.)	$\times 10^4/\mu l$	32.6 ± 9.6	

Abbreviation
Hb: hemoglobin
Ht: hematocrit
MCH: mean corpuscular hemoglobin
MCHC: mean corpuscular hemoglobin concentration
MCV: mean corpuscular volume
RBC: red blood cell
WBC: white blood cell

5.9 Bleeding System for Obtaining More Correct Physiological Data and Animal Welfare

Laboratory rabbits are traditionally kept individually in small cages with restricted or free (*ad libitium*) commercial diets. This has led to several physiological problems related to the fact that they move too little, as well as behavioral disorders. Over the past 20 years many laboratories have improved the cages for rabbits (sizes and materials), but also by introducing floor pens and group keeping. However, there are still some problems because they can be very aggressive when kept in groups.

Laboratory rabbits are housed singly in small cages during the research. The cages for laboratory rabbits have changed from wood to metal and now plastic caging. Keeping in small cages can lead to several abnormal behaviors and to reduced welfare of rabbits.

Group keeping of rabbits in floor pens has been introduced in many laboratories and countries during the last 20 years. This is beneficial for the rabbits because

Table 5.8 Values of serum biochemical and enzyme parameters of NZW

		♂	♀
Total protein	g/dl	6.90 ± 0.36	6.70 ± 0.41
Albumin	g/dl	3.39 ± 0.29	3.04 ± 0.26
α1-gobulin	g/dl	0.60 ± 0.12	0.37 ± 0.08
α2-gobulin	g/dl	0.43 ± 0.09	0.21 ± 0.05
β-gobulin	g/dl	1.0 ± 10.20	1.46 ± 0.24
γ-gobulin	g/dl	1.46 ± 0.21	1.69 ± 0.22
A/G ratio		0.97 ± 0.16	0.8 ± 30.15
Glucose	mg/dl	135 ± 12.0	128 ± 14.0
Urea nitrogen	mg/dl	19.2 ± 4.93	17.6 ± 4.36
Uric acid	mg/dl	2.65 ± 0.88	2.62 ± 0.87
Amylase	Somogy IU/l	132 ± 16.0	127 ± 12.0
Bililbin	mg/dl	0.32 ± 0.04	0.30 ± 0.04
Cholesterol	mg/dl	26.7 ± 12.9	24.5 ± 11.2
Triglycerides[a]	mg/dl	74 ± 38.0	80 ± 40.8
Creatinin	mg/dl	1.59 ± 0.34	1.67 ± 0.38
GPT	IU/l	65.7 ± 6.54	62.5 ± 5.85
GOT	IU/l	72.3 ± 12.0	68.1 ± 10.5
LDH	IU/l	84.4 ± 20.5	78.5 ± 22.0
CK	IU/l	1.35 ± 0.56	1.30 ± 0.45
ALP	IU/l	10.4 ± 2.28	9.96 ± 3.10
AcP	IU/l	1.56 ± 0.53	1.40 ± 0.38
Na	mEq/l	146 ± 1.15	141 ± 1.40
K	mEq/l	5.75 ± 0.20	6.40 ± 0.16
Cl	mEq/l	101 ± 1.45	105 ± 1.22
Ca	mg/dl	10.0 ± 1.11	9.50 ± 1.10
P	mg/dl	4.82 ± 1.05	5.06 ± 0.93
Mg	mg/dl	2.52 ± 0.24	3.20 ± 0.22

[a][24]

Abbreviation
AcP: acid phoshatase
ALP: alkaline phoshatase
CK: creatinin kinase
GOT: glutamate oxaloacetate transaminase
GPT: glutamate pyrurate transaminase
LDH: lactate dehaydrogenase

they can express social behaviors and exercise (Batchelor 1991; Heath and Stott 1990). Abnormal behaviors and physiological conditions caused by small environments are reduced. But the risk of fighting between rabbits is increased (Morton et al. 1993). When we make a group keeping of rabbits, we have to consider the following points: the compatibility of individual animals, the size of pens, the keeping density, the husbandry technique and environmental enrichment (Morton et al. 1993).

References

Altman PL and Dittmer DS. (1964) Body weight: Mammals other than man and rodents. In "Biology Data Book": 99–100. Federation of American Societies for Experimental Biology, Washington, DC.

Barzago MM, Botolotti A, Omarini D, et al. (1992) Monitoring of blood gas parameters and acid-base balance of pregnant and non-pregnant rabbit (Oryctolagus cuniculus) in routine experimental conditions. Lab. Anim. 26: 73–78.

Batchelor GR (1991) Group housing on floor pens and enrichment in sandy lop rabbits. Anim. Technol. 42: 109–120.

Batchelor GR (1999) The laboratory rabbit. In "The UFAW Handbook on the Care and Management of Laboratory Animals" 7th ed.: 395–408. Blackwell Science, Oxford.

Bell DJ (1999) The European wild rabbit. In "The UFAW Handbook on the Care and Management of Laboratory Animals" 7th ed.: 389–394. Blackwell Science, Oxford.

Cheeke PR and Amberg JW (1973) Comparative calcium excretion by rats and rabbits. J. Anim. Sci. 37: 450–454.

Cox RH (1978) Comparison of carotid artery mechanics in the rat, rabbits, and dog. Am. J. Physiol. 234: H280–H288.

Flatt RE and Carpenter AB (1971) Identification of crystalline material in urine of rabbit. Am. J. Vet. Res. 32: 655–658.

Gillett CS (1994) Selected drug dosages and clinical reference data. In "The Biology of the Laboratory Rabbit" 2nd ed.: 467–472. Academic, San Diego, CA.

Gonzalez RR, Kluger MJ and Stolwijk AJ (1974) Thermoregulatory responses to thermal transients in the rabbit. Am. J. Physiol. 227: 1292–1298.

Hagen KW (1974) Colony husbandry. In "The Biology of the Laboratory Rabbit": 23–47. Academic, New York.

Harkness JE and Wagner JE (1983) "The Biology and Medicine of rabbits and rodents." 2nd ed. Lea & Febiger, Philadelphia, PA.

Heath M and Stott E (1990) Housing rabbits the unconventional way. Anim. Technol. 41: 13–25.

Hennemann PH (1959) Effects of high calcium intakes on renal function. Fed. Proc. 18: 1093–1096.

Kaplan HM and Timmons EH (1979) "The Rabbit – A Model for the Principles of Mammalian Physiology and Surgery." Academic, New York.

Kersten AM, Meijsser FM, Metz JH (1989) Effects of early handling on later open-field behavior in rabbits. Appl. Anim. Behav. Sci. 24: 157–167.

Matsuki T, Cohen MV and Downey JM (1990) Free radical scavengers preserve wall motion in the xanthine oxidase-deficient rabbit heart. Coron. Artery Dis. 1: 383–390.

Meredith A (2000) General biology and husbandry. In "Manual of Rabbit Medicine and Surgery": 13–23. British Small Animal Veterinary Association, London.

Mitruka BM and Rawnsley HM (1977) "Clinical Biochemical and Hematological Reference Value in Normal Experimental Animals." Masson, New York.

Morton DB, Jennings M, Batchelor GR, et al. (1993) Refinement in rabbit husbandry. Lab. Anim. 27: 301–329.

Myers K, Parer I, Wood D, et al. (1994) The rabbit in Australia. In "The European Rabbit: The History and Biology: The History and Biology of a Successful Coloniser": 108–157. Oxford University Press, Oxford.

Mykytowycz R (1968) Territorial marking by rabbits. Sci. Am. 218: 116–126.

Sanford TD and Colby ED (1980) Effect of xylazine and ketamine on blood pressure, heart rate, and respiratory rate in rabbits. Lab. Anim. Sci. 30: 519–523.

Suckow MA, Brammer DW, Rush HG and Chrisp CE (2002) Biology and diseases of rabbits. In "Laboratory Animal Medicine" 2nd ed.: 329–364. Academic, San Diego, CA.

Wolford ST, Schroer RA, Gohs FX, et al. (1986) Reference range data base for serum chemistry and hematology values in laboratory animals. J. Toxicol. Environ. Health 18: 161–188.
Yokoyama E (1979) Flow-volume curves of excised right and left rabbit lungs. J. Appl. Physiol. 46: 461–468.
Zarrow MX, Denenberg VH, Anderson CO (1965) Rabbit: Frequency of suckling in the pup. Science 150: 1835–1836.

Chapter 6
Rabbit Transgenesis

Shuji Kitajima, Enqi Liu, and Jianglin Fan

Abstract In this chapter, we will introduce the methods for the generation of transgenic rabbits. These methods are those used for transgenic mice but with some modifications and refinements. We will review all possible methods for rabbit transgenesis with emphasis on microinjection method because it is the most popular method so far.

Keywords Embryo transfer, Pronuclear microinjection, Superovulation, Transgenic rabbits

6.1 Introduction

After the first report of transgenic mice introduced a foreign gene in their chromosome by Gordon et al. (1980) in 1980, transgenic animals including rabbits have been widely generated in the world. Recently, the value of transgenic animals such as human disease models and bioreactors for producing therapeutic proteins in the milk has been increasing more and more in the field of medical science and drug industry. In general, genetically modified animals can be classified in transgenic and knockout animals. Transgenic animals refer to animals with integrated foreign gene into their chromosome and show "gain of function". On the other hand, knockout animals are those with their endogenous gene disrupted by homologous recombination using embryonic stem (ES) cell and show "loss of function".

S. Kitajima (✉)
Division of Biological Resources and Development, Analytical Research Center for Experimental Sciences, Saga University, 5-1-1 Nabeshima, Saga 849-8501, Japan
e-mail: kitajims@cc.saga-u.ac.jp

E. Liu
Experimental Animal Center, Xi'an Jiaotong University, Xi'an, China

J. Fan
Department of Molecular Pathology, University of Yamanashi, Yamanashi, Japan

L.-M. Houdebine and J. Fan (eds.), *Rabbit Biotechnology: Rabbit Genomics, Transgenesis, Cloning and Models*,

Until now, it has been reported several methods for generating transgenic animals such as pronuclear microinjection, gene targeting using ES cells, retroviral vector, sperm vector and somatic cell nuclear transfer (SCNT) technique. For generation of transgenic rabbits, the most successful commonly used method is pronuclear microinjection, while use of the other methods is still restricted to mice and not been fully established for practical use. In this chapter, we introduce the protocol of microinjection method currently used in our laboratory. Readers are also encouraged to refer to the original review articles published Bosze et al. (2003), Fan et al. (1999), Fan and Watanabe (2003) and Kitajima et al. (2003).

6.2 Pronuclear Microinjection Methods for Rabbit Transgenesis

The first transgenic rabbit was reported by Hammer et al. (1985) using microinjection method in 1985. By microinjection method, foreign gene (called DNA construct) is directly injected into the pronucleus of fertilized egg using fine glass pipette under microscope (Fig. 6.1). Then, embryos are transplanted to synchronized-induced ovulated foster mothers to make them develop to individuals. We can expect a part of pups born after embryo transfer to bear the foreign gene (called transgene) integrated into their chromosome, but the gene-transfer rate is low which can be usually less than 5% of total pups.

The production efficiency of transgenic rabbit is thought to be influenced by several factors such as quality of DNA quality (concentration, purity and size etc.) used for microinjection along with skills of technicians. Common problems are low pregnancy rate, small size of pups after embryo transfer, and low gene-transfer rate. In general, the pregnancy rate after embryo transfer was nearly 50% and the number of newborn pups was 10% or lower for transferred embryos. The gene-transfer rate was 5% or lower for the pups obtained by embryo transfer.

6.3 Basic Technique for Microinjection

6.3.1 Superovulation

In order to collect as many embryos as possible from one donor rabbit, superovulation is generally induced by several hormone treatments. In rabbits, pregnant mare's serum gonadotropin (PMSG) (Fan et al. 1999; Keefer et al. 1985; Zheng et al. 2004) or follicular stimulating hormone (FSH) (Al-Hasani et al. 1986; Cheng et al. 1999; Hashimoto et al. 2004; Hashimoto et al. 2007; Kauffman et al. 1998) have been commonly used for superovulation. Superovulation protocol used in our laboratory is shown in Fig. 6.2. In the case of PMSG, on the first day donor rabbits are intramuscularly single injected with 150U of PMSG. In the case of FSH, 0.5 AU of hormone is subcutaneously administered in 12-h intervals for 3 days, totally

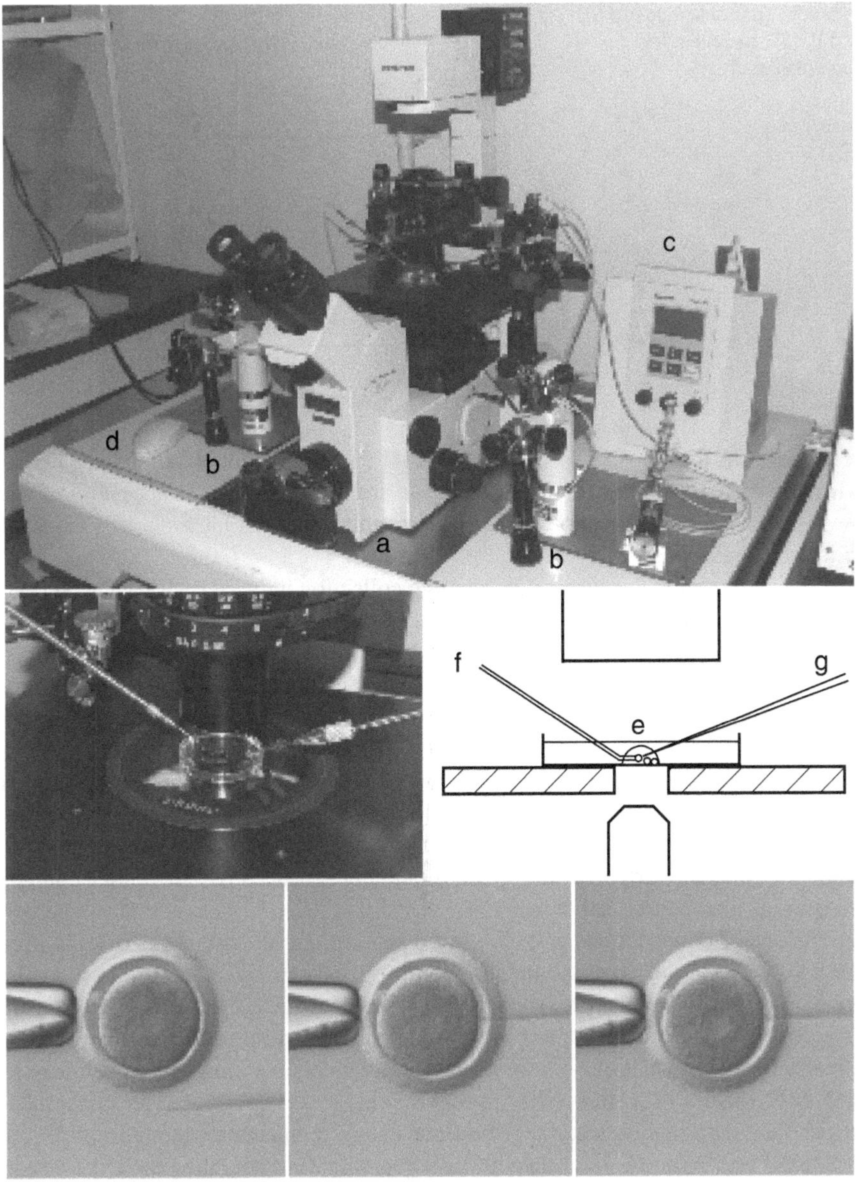

Fig. 6.1 Basic equipments for pronuclear microinjection, *Top*: Invented microscope (**a**) with a differential interference device is equipped with micromanipulators (**b**), microinjector (**c**) and its controller (**d**). *Middle*: The left photograph is a magnified composition of the microscope stage and the right picture is a scheme viewed from the side. The dish is placed on stage of microscope, which is warmed by a thermo plate at 37°C during microinjection. The drop of medium containing embryos (**e**) is covered by mineral oil in the dish. For microinjection, a fertilized egg at the pronuclear stage is hold with a "holding pipette" (**f**) by operating micromanipulator (**b** in upper photograph). "Injection pipette" (**g**) is located right side of the stage and it is connected to the microinjector (FemtoJet, Eppendorf, Germany). *Bottom*: Phographs from left to right shows the process of pronuclear microinjection. After microinjection, the enlargement of the male pronucleus (right photograph) can be observed

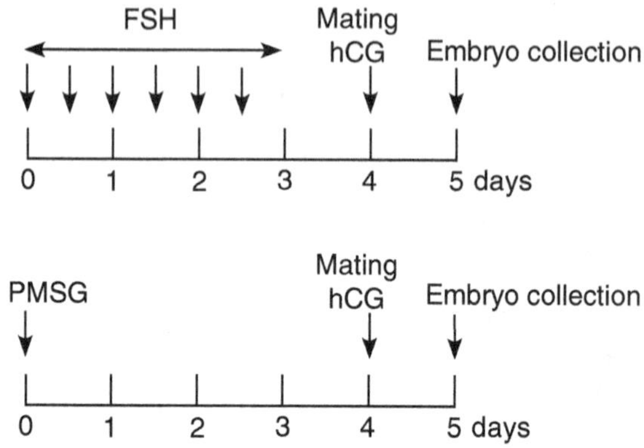

Fig. 6.2 Time schedule of FSH and PMSG treatment for superovulation

six times. On the fourth day (both of PMSG and FSH), donor rabbits are mated with two different males to ensure the eggs are fertilized in our laboratory. After mating, 100 U of human chorionic gonadotrophin (hCG) is intravenously injected to induce ovulation.

When comparing the performance of egg collection induced by FSH and PMSG, we are able to obtain a more stable number of eggs by FSH than by PMSG. Based on our data in the past 5 years, the average of number of embryos obtained from FSH- and PMSG-induced Japanese White (JW) donor rabbit is 41.1 and 28.6, respectively. The PMSG treatment is not consistent because we sometimes observe that rabbits showed either no response (no eggs) or an excessive response (more than 100 fertilized eggs per rabbit) to PMSG injection. At any rate, fertilized eggs obtained from both hormone treatments are usable for microinjection. In our experience, it seems that there is no apparent difference between these two methods in terms of in vitro embryo development, the number of pups obtained after embryo transfer, or transgenesis efficiency.

The age of donor rabbits and the season also influence the number of eggs irrespectively of the hormone treatment. We have used the JW rabbits older than 16 weeks old for superovulation. Animals of 16 weeks of age are old enough to be used as a donors with both FSH and PMSG treatments. We also used 16–40 weeks old females for superovualtion, and we observed that younger animals are as sensitive to both of FSH and PMSG than older ones. As far as seasonal variations are concerned, in our facilities, rabbits were kept under constant temperature, humidity (23°C ± 2°C, 55% ± 15%) and light cycle with 12-h interval throughout the year. Under these conditions, we did not find any seasonal variation in embryo collection.

6.3.2 Embryo Collection

Fertilized eggs are recovered by flushing through the oviducts by appropriate medium such as 10% FBS-PB1 and M2 medium. For microinjection, we collect the embryos 17–19 h after mating (injection of hCG). In our laboratory, we remove

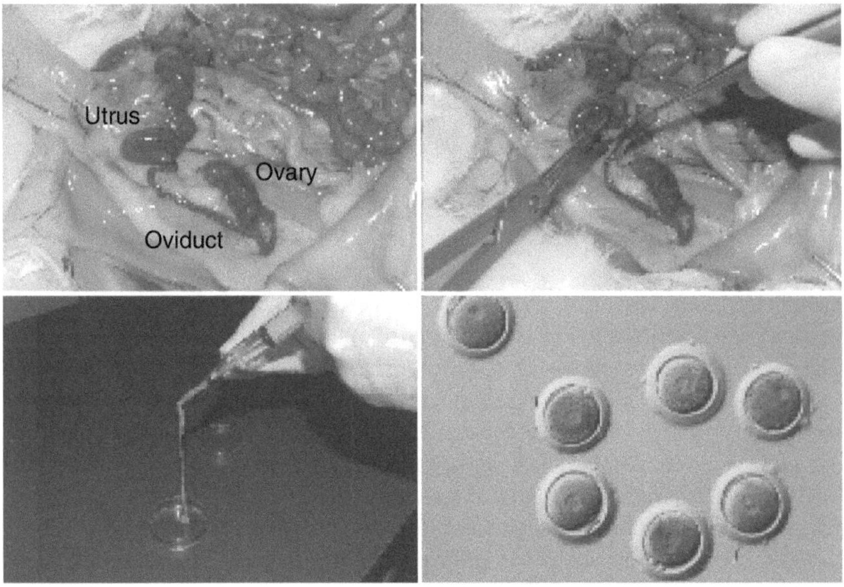

Fig. 6.3 Embryo collection. *Top*: For microinjection, we collect embryos 17–19 h after mating (injection of hCG). In our laboratory, we remove the oviduct tube and a part of uterus from donor rabbit after the sacrifice by overdose of an appropriate anesthetic. *Bottom*: Oviduct is flushed by medium from uterus to fimbriae (left photo) and collect flushing medium containing feritilized eggs (right photo) into dish. Egg in right photograph is at the pronuclear stage and it was collected 17 h after mating (injection of hCG). We can see two pronuclei, male and female pronuclei meeting at the center of the egg cytoplasm

the oviduct tube and a part of uterus from donor rabbit after sacrifice by an overdose of appropriate anesthetic. The fertilized eggs are in ampulla 17–19 h after injection of hCG. After removing the oviduct, we insert 18 gage needle connected to 10 ml syringe from uterus side. Then oviduct is flushed with each 5–10 ml of medium for one side and flushing medium is collected into a test tube or a dish (Fig. 6.3).

6.3.3 Pronuclear Microinjection

In our laboratory, pronuclear microinjection is done with embryos 19–21 h after mating. DNA solution is injected into a pronucleus using micromanipulators and a microinjector under the inverted microscope equipped with a differential interference device (Fig. 6.1). We use DNA solution at the concentration of 3–5 ng/ml for microinjection. In general, the male pronuclei are larger than female ones, so DNA solution is usually injected into the male pronucleus. When the tip of injection pipette has reached the pronucleus, the position of the injection pipette is kept and DNA solution is injected until the swelling of the pronucleus (Fig. 6.1). For the microinjection, purity and concentration of DNA solution are important.

When purity of DNA solution is low and contains many contaminants, injection pipette will be clogged up immediately and microinjection cannot be pursued. In addition, when concentration of DNA solution is too high, the in vitro development rate of embryos after microinjection is lower.

6.3.4 Embryo Transfer

One specific characteristic of a rabbit embryo is a thick mucin layer that forms around the zona pellucida in the oviduct. The presence of this mucin layer greatly influences pregnancy. If there is no mucin layer, the pregnant rate will drop dramatically (Moore et al. 1968). Thus, it is desirable to quickly transfer the embryo into the recipient rabbits after microinjection in order to allow the mucin layer formation around the zona pellucida. Recipient rabbits can be prepared by injection with 50U of hCG at same time when donor was mated and injected with hCG. In our laboratory, embryos are cultured for 2–3h in a CO_2 incubator (5% CO_2, 38.5°C) after microinjection, and 20–30 of survived embryos are transferred to each recipient. The survival rate of embryos after cultivation for 2–3h is between 80–90%. It can be strongly influenced by technician's skill if there is no problem with quality of DNA solution.

For embryo transfer, recipient rabbits should be kept under appropriate anesthesia. We inject intravenously the mixture of ketamine and medetomidine (ketamine 25mg and medetomidine 5mg in 1ml) to a recipient weighted about 3kg. This anesthesia will be effective for 30–45min a time sufficient for the embryo transfer. Under anesthesia, the adipose tissue surrounding the ampulla is pulled out of the body cavity by flank incision (Fig. 6.4). Then, embryos are transferred from fimbriae into oviduct with a small amount of medium (about 10–20ml) using indwelling needle (18G) connected to microsyringe filled with saline (Fig. 6.4). By this method, the pregnancy rate over a 5 years period was 57.1–77.8% (an average was 66.2%), the ratio of pups for the number of transferred embryos was 5.6–15.8% (an average was 9.0%).

To increase the yield of transgenic rabbits, it is recommended to use multiparous does which had shown good nursing before being used as recipients. In addition, it can be preferable to prepare another foster mother mated with a fertile male 1 or 2 days before the mating of donor rabbits. Indeed, when the number of fetuses of the recipient female is only 1 or 2, it often does not give birth and Caesarean section is thus sometimes needed. The pups born by Caesarean section can then be added to those of the foster mother with a higher chance to develop normally.

6.3.5 Detection of Transgene in Newborns

We can do the screening of transgenic pups born after embryo transfer by polymerase chain reaction (PCR) or Southern blotting analysis. In generally, genomic DNA can be collected from ear biopsy or blood sample when pups are 4 weeks old. If the

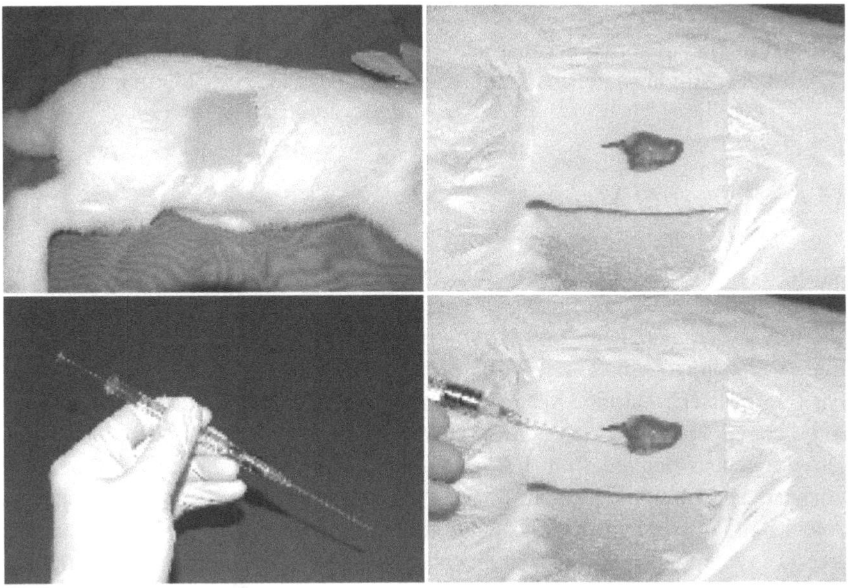

Fig. 6.4 Embryo transfer. *Top*: Recipient rabbit should be placed under appropriate anesthesia (left photograph). Under anesthesia, the adipose tissue surrounding the ampulla is pulled out of the body cavity by flank incision (right photograph). *Bottom*: For embryo transfer, we use a microsyringe connected indwelling needle (18G) and filled it with saline (left photograph). Then, embryos are transferred from fimbriae into oviduct with small amount of medium (about 10–20 ml) (right photograph)

proteins derived from expression of transgene are secreted in blood or milk, they can directly detected by Western blotting analysis or enzyme-linked immunosorbent assay (ELISA) using specific antibodies. In our data over 5 years, the efficiency of transgenesis was 1.5–16.7% of pups (average 6.4%) obtained after embryo transfer. Generation of F1 animals obtained after mating F0 with wild rabbits is needed to know whether the F0 transmit the transgene to progeny and express it in specific organs.

6.4 Other Possible Future Methods for Rabbit Transgenesis

Pronuclear microinjection is the most popular method for the production of a number of transgenic animals although it is not efficient at all in some species and it is used routinely in the world. However, it cannot control the copy number and the integrating sites of transgenes into the host chromosome. This problem results in the so-called position-effects leading to the failure of specific expression of the transgene. Moreover, the foreign gene may inactivate an endogenous gene into which it randomly integrated. Currently, some research groups are developing new methods instead of the pronuclear microinjection. For example, retroviral vector, sperm vector, establishment of rabbit ES cells for gene targeting and SCNT technology are expected to become efficient alternative methods to produce genetically modified rabbits, but it still needs further studies to adapt these methods to rabbits.

6.4.1 Retroviral Vectors

Retroviral genomes are known to integrate into host chromosomes after infection and this natural mechanism makes it possible the generation of viral vectors able to introduce exogenous genes into host chromosome for the production of transgenic animals. Recently, Dull et al. (1998) developed a packaging system making it available very efficient lentiviral vectors for transgenesis. In this system, a recombinant lentiviral genome in virus particles produced by packaging cells contains backbone DNA sequences derived from human immunodeficiency virus type I, promoters, enhancers and foreign genes but none of the deleterious viral genes. The envelope protein is the G protein of a vesicular stomatitis virus which has the capacity to target the recombinant particles in all cell types. Moreover, the promoter and a part of the enhancer region of long terminal repeat U3 are deleted generating a safe self-inactivating recombinant virus able to infect only once targeted cells but not to disseminate. Until now, several transgenic animals have been produced using recombinant lentiviral vectors (Hofmann et al. 2004; Hofmann et al. 2003; Lois et al. 2002; Mcgrew et al. 2004; Pfeifer et al. 2002; Van Den Brandt et al. 2004). But, successful production of transgenic rabbits using retroviral vector has not been reported yet.

6.4.2 Sperm Vector

In 1971, Brackett et al. (1971) reported that rabbit sperm could bring a foreign gene to egg at fertilization. In 1989, Lavitrano et al. (1989) produced transgenic mice by adding exogenous foreign gene to the medium in which sperm was incubated before in vitro fertilization (IVF) was achieved. As the result, they found the occurrence of transgenesis in 30% of offspring after embryo transfer. More recently, the production of transgenic rabbit using sperm-mediated gene transfer (SMGT) method was reported (Li et al. 2006; Shen et al. 2006). Shen et al. (Shen et al. 2006) generated transgenic rabbit by mating with a male rabbit in which an exogenous gene had been injected into their testis. This method is known as testis-mediated gene transfer (TMGT). The authors reported that 48.4% of pups were transgenic and found the expression of the transgene in 14 of 17 pups. However, unfortunately the detail of the protocol was not clearly described and the authors could not observe any transmission of the transgene into offspring after mating founder transgenic rabbits (Li et al. 2006; Shen et al. 2006). SMGT method may be very simple, highly efficient and of a low cost in comparison with the conventional microinjection (Smith and Spadafora 2005). The SMGT method must be dramatically improved before its efficiency makes it possible the production of transgenic rabbits. Further development of this method is expected.

6.4.3 Gene Targeting Using ES Cells

No rabbit ES cell lines capable of transmitting a genetic modification through germ line to progeny have been fully established yet. If rabbit ES cell lines were established, they would be very useful to produce genetically modified rabbit, not only transgenics but also knockout rabbits by the gene targeting method. The methods to establish rabbit ES cells are described in another chapter of this book elsewhere. Development of rabbit ES cells is expected in the future.

6.4.4 Somatic Cell Nuclear Transfer (SCNT)

Success of cloning in mammals by SCNT began with the birth of "Dolly" the sheep, and the method was extended to bovine, laboratory mice, goats, pigs and a few other species (Baguisi et al. 1999; Cibelli et al. 1998; Wakayama et al. 1998; Wilmut et al. 1997). First cloned rabbit was reported by Chesne et al. (2002) in 2002. These authors obtained six rabbit clones by transplanting reconstituted embryos using cumulus cells as nuclear donors. Cloning efficiency reported to the number of the reconstituted embryos was 1.6%. The current efficiency to generate rabbit clones for transgenesis is not high in comparison to the microinjection method. But SNCT could become helpful for the production of the genetically modified rabbits if the cells used as nuclear donors have been previously genetically modified in vitro. This includes random gene addition and gene knockout using targeted gene replacement by homologous recombination. In fact, transgenic clones were produced by SNCT technique in cow, sheep and swine (Cibelli et al. 1998; Lai et al. 2002; Mccreath et al. 2000). This method may be adapted for producing transgenic rabbit clones. However, the techniques for producing animal clones by SCNT remains difficult and it is not generally spread over the world.

References

Al-Hasani, S., Tolksdorf, A., Diedrich, K., Van Der Ven, H. and Krebs, D. (1986) Successful in-vitro fertilization of frozen-thawed rabbit oocytes. Hum Reprod 1:309–312.

Baguisi, A., Behboodi, E., Melican, D.T., Pollock, J.S., Destrempes, M.M., Cammuso, C., Williams, J.L., Nims, S.D., Porter, C.A., Midura, P., Palacios, M.J., Ayres, S.L., Denniston, R.S., Hayes, M.L., Ziomek, C.A., Meade, H.M., Godke, R.A., Gavin, W.G., Overstrom, E.W. and Echelard, Y. (1999) Production of goats by somatic cell nuclear transfer. Nat Biotechnol 17:456–461.

Bosze, Z., Hiripi, L., Carnwath, J.W. and Niemann, H. (2003) The transgenic rabbit as model for human diseases and as a source of biologically active recombinant proteins. Transgenic Res 12:541–553.

Brackett, B.G., Baranska, W., Sawicki, W. and Koprowski, H. (1971) Uptake of heterologous genome by mammalian spermatozoa and its transfer to ova through fertilization. Proc Natl Acad Sci U S A 68:353–357.

Cheng, H., Dooley, M.P., Hopkins, S.M., Anderson, L.L., Yibchok-Anun, S. and Hsu, W.H. (1999) Development of rabbit embryos during a 96-h period of in vitro culture after superovulatory treatment under conditions of elevated ambient temperature. Anim Reprod Sci 56:279–290.

Chesne, P., Adenot, P.G., Viglietta, C., Baratte, M., Boulanger, L. and Renard, J.P. (2002) Cloned rabbits produced by nuclear transfer from adult somatic cells. Nat Biotechnol 20:366–369.

Cibelli, J.B., Stice, S.L., Golueke, P.J., Kane, J.J., Jerry, J., Blackwell, C., Ponce De Leon, F.A. and Robl, J.M. (1998) Cloned transgenic calves produced from nonquiescent fetal fibroblasts. Science 280:1256–1258.

Dull, T., Zufferey, R., Kelly, M., Mandel, R.J., Nguyen, M., Trono, D. and Naldini, L. (1998) A third-generation lentivirus vector with a conditional packaging system. J Virol 72:8463–8471.

Fan, J. and Watanabe, T. (2003) Transgenic rabbits as therapeutic protein bioreactors and human disease models. Pharmacol Ther 99:261–282.

Fan, J., Challah, M. and Watanabe, T. (1999) Transgenic rabbit models for biomedical research: current status, basic methods and future perspectives. Pathol Int 49:583–594.

Gordon, J.W., Scangos, G.A., Plotkin, D.J., Barbosa, J.A. and Ruddle, F.H. (1980) Genetic transformation of mouse embryos by microinjection of purified DNA. Proc Natl Acad Sci U S A 77:7380–7384.

Hammer, R.E., Pursel, V.G., Rexroad, C.E., Jr., Wall, R.J., Bolt, D.J., Ebert, K.M., Palmiter, R.D. and Brinster, R.L. (1985) Production of transgenic rabbits, sheep and pigs by microinjection. Nature 315:680–683.

Hashimoto, S., Kuramochi, T., Aoyagi, K., Takahashi, R., Ueda, M., Hirao, M., Kamei, M., Kitada, K. and Hirasawa, K. (2004) Refined porcine follicle stimulating hormone promotes the responsiveness of rabbits to multiple-ovulation treatment. Exp Anim 53:395–397.

Hashimoto, S., Kimura, K., Kuramochi, T., Aoyagi, K., Hirako, M., Kawaguchi, M., Iwata, H., Hirao, M., Kitada, K., Hirasawa, K. and Ueda, M. (2007) Responsiveness of rabbits to superovulation treatment by a single injection of follicle-stimulating hormone with aluminum hydroxide gel. Mol Reprod Dev 74:1208–1212.

Hofmann, A., Kessler, B., Ewerling, S., Weppert, M., Vogg, B., Ludwig, H., Stojkovic, M., Boelhauve, M., Brem, G., Wolf, E. and Pfeifer, A. (2003) Efficient transgenesis in farm animals by lentiviral vectors. EMBO Rep 4:1054–1060.

Hofmann, A., Zakhartchenko, V., Weppert, M., Sebald, H., Wenigerkind, H., Brem, G., Wolf, E. and Pfeifer, A. (2004) Generation of transgenic cattle by lentiviral gene transfer into oocytes. Biol Reprod 71:405–409.

Kauffman, R.D., Schmidt, P.M., Rall, W.F. and Hoeg, J.M. (1998) Superovulation of rabbits with FSH alters in vivo development of vitrified morulae. Theriogenology 50:1081–1092.

Keefer, C.L., Bennett, K.A. and Brackett, B.G. (1985) In vitro fertilization in the rabbit after delayed ovum recovery. Biol Reprod 33:388–392.

Kitajima, S., Morimoto, M., Watanabe, T. and Fan, J. (2003) Transgenic Rabbits: Production and Application for Human Disease Models. Recent Res Devel Biophys Biochem 3:179–194.

Lai, L., Kolber-Simonds, D., Park, K.W., Cheong, H.T., Greenstein, J.L., Im, G.S., Samuel, M., Bonk, A., Rieke, A., Day, B.N., Murphy, C.N., Carter, D.B., Hawley, R.J. and Prather, R.S. (2002) Production of alpha-1,3-galactosyltransferase knockout pigs by nuclear transfer cloning. Science 295:1089–1092.

Lavitrano, M., Camaioni, A., Fazio, V.M., Dolci, S., Farace, M.G. and Spadafora, C. (1989) Sperm cells as vectors for introducing foreign DNA into eggs: genetic transformation of mice. Cell 57:717–723.

Li, L., Shen, W., Min, L., Dong, H., Sun, Y. and Pan, Q. (2006) Human lactoferrin transgenic rabbits produced efficiently using dimethylsulfoxide-sperm-mediated gene transfer. Reprod Fertil Dev 18:689–695.

Lois, C., Hong, E.J., Pease, S., Brown, E.J. and Baltimore, D. (2002) Germline transmission and tissue-specific expression of transgenes delivered by lentiviral vectors. Science 295:868–872.

Mccreath, K.J., Howcroft, J., Campbell, K.H., Colman, A., Schnieke, A.E. and Kind, A.J. (2000) Production of gene-targeted sheep by nuclear transfer from cultured somatic cells. Nature 405:1066–1069.

Mcgrew, M.J., Sherman, A., Ellard, F.M., Lillico, S.G., Gilhooley, H.J., Kingsman, A.J., Mitrophanous, K.A. and Sang, H. (2004) Efficient production of germline transgenic chickens using lentiviral vectors. EMBO Rep 5:728–733.

Moore, N.W., Adams, C.E. and Rowson, L.E. (1968) Developmental potential of single blastomeres of the rabbit egg. J Reprod Fertil 17:527–531.

Pfeifer, A., Ikawa, M., Dayn, Y. and Verma, I.M. (2002) Transgenesis by lentiviral vectors: lack of gene silencing in mammalian embryonic stem cells and preimplantation embryos. Proc Natl Acad Sci USA 99:2140–2145.

Shen, W., Li, L., Pan, Q., Min, L., Dong, H. and Deng, J. (2006) Efficient and simple production of transgenic mice and rabbits using the new DMSO-sperm mediated exogenous DNA transfer method. Mol Reprod Dev 73:589–594.

Smith, K. and Spadafora, C. (2005) Sperm-mediated gene transfer: applications and implications. Bioessays 27:551–562.

Van Den Brandt, J., Wang, D., Kwon, S.H., Heinkelein, M. and Reichardt, H.M. (2004) Lentivirally generated eGFP-transgenic rats allow efficient cell tracking in vivo. Genesis 39:94–99.

Wakayama, T., Perry, A.C., Zuccotti, M., Johnson, K.R. and Yanagimachi, R. (1998) Full-term development of mice from enucleated oocytes injected with cumulus cell nuclei. Nature 394:369–374.

Wilmut, I., Schnieke, A.E., Mcwhir, J., Kind, A.J. and Campbell, K.H. (1997) Viable offspring derived from fetal and adult mammalian cells. Nature 385:810–813.

Zheng, Y.L., Jiang, M.X., Zhang, Y.L., Sun, Q.Y. and Chen, D.Y. (2004) Effects of oocyte age, cumulus cells and injection methods on in vitro development of intracytoplasmic sperm injection rabbit embryos. Zygote 12:75–80.

Chapter 7
Rabbit as a Model for the Study of Human Diseases

Masashi Shiomi

Abstract Although genetically modified mice are playing an essential role in the study of the expression and functions of individual genes, rabbits are useful animal models to extrapolate animal studies to humans. It is necessary that key gene expression and function are equivalent and close to human rather than the outward features or phenotype. For example, to study human hypercholesterolemia, the only hypercholesterolemia is insufficient, the lipoprotein profiles and enzymes in the lipoprotein metabolism of animal models are important for translational medicine. Lipoprotein metabolism of rabbits resembles humans closely. In addition, histopathological and/or immunohistochemical features of the tissues of disease similar to humans are important. In this field, spontaneous hypercholesterolemic rabbits (WHHL and WHHLMI rabbits) have contributed to the elucidation of lipoprotein metabolism, atherogenesis, and to the development of therapeutic compounds, such as statins. Recently, a number of transgenic rabbits have been developed and they also contribute to the study of cardiac function and infectious diseases. Furthermore, rabbits are useful for studies of orthopedic surgery, cardiovascular surgery, and neoplastic diseases. Rabbit models have contributed not only to the mechanistic studies of human diseases but also to the development of therapeutic compounds, devices, or techniques for therapeutics. Applying these animal models in translational researches promotes the elucidations of human diseases.

Keywords animal models for human diseases, translational research, transgenic rabbits, WHHL/WHHLMI rabbits

7.1 Introduction

After the genomes of human and mouse were fully deciphered, it has been recognized that the analyses of gene expression and functions are important to understand the pathogenesis and the mechanisms of diseases. It is critical for researchers to choose

M. Shiomi (✉)
Institute for Experimental Animals, Kobe University School of Medicine, 7-5-1, Kusunoki-cho, Chuo-ku, Kobe 650-0017, Japan
e-mail: ieakusm@med.kone-u.ac.jp

L.-M. Houdebine and J. Fan (eds.), *Rabbit Biotechnology: Rabbit Genomics, Transgenesis, Cloning and Models*,

appropriate animal species or animal models to mimic human physiology so that the results derived from experimental animals can be translated from "bench to bed" for the development of diagnostic and therapeutic strategies. In this chapter, I introduce rabbit models for their use in human disease research.

7.2 Studies Using Rabbit Models

As shown in Table 7.1, rabbit models have been widely used for various purposes. Rabbits were initially used for the production of antibodies and non-clinical safety studies for testing compounds (ocular tolerance test, dermal tolerance test, pyrogen

Table 7.1 Rabbit models for human diseases

Human diseases	Rabbit models
Lipid metabolism disorder	
Hypercholesterolemia	WHHL rabbit, WHHLMI rabbit, cholesterol/western diet-fed rabbit
Hyperlipidemia	SMHL rabbit
Postprandial hypertriglyceridemia	PHT rabbit
Metabolic syndrome	WHHLMI rabbit, SMHL rabbit, PHT rabbit
Cardiovascular diseases	
Atherosclerosis	WHHL rabbit, WHHLMI rabbit, cholesterol/western diet-fed ordinary rabbit
Coronary atherosclerosis	WHHL rabbit, WHHLMI rabbit, cholesterol/western diet-fed ordinary rabbits
Cerebral atherosclerosis	WHHL rabbit, WHHLMI rabbit
Myocardial infarction	WHHLMI rabbit
Cardiac function	WHHLMI rabbit, ordinary normal rabbits
Prolonged QT syndrome	Transgenic rabbits
Other diseases	
Xanthoma	WHHL rabbit, WHHLMI rabbit
Infectious diseases	New Zealand white rabbits, transgenic rabbits (Tuberculosis, Papillomavirous, Prion diseases)
Inflammatory response	Transgenic rabbits
Orthopedic surgery	Ordinary normal rabbits
	(Osteonecrosis, meniscal repair, autologous osteochondral transplantation, osteochondral defects)
Cardiovascular surgery	Ordinary normal rabbits
	(Spinal cord ischemia due to back-bleeding, dysfunction of aortic valve)
Transplantable Neoplasms	Ordinary normal rabbits
	(VX2 lung cancer, Kato sarcoma, Kondrateva osteogenic sarcoma, Brown-Pearce Carcinoma)
Keratopathy	WHHL rabbit, WHHLMI rabbit
Overactive bladder	WHHLMI rabbit
Hypacusia	WHHL rabbit
Chronic pancreatitis	WHHL rabbit

test, and reproduction toxicity study) and so on. From 1980, rabbit models for human diseases have been developed either from naturally mutant animals, or by selective breeding or genetic manipulations. These rabbit models have been used for studies of human diseases as described below.

7.3 Rabbit Models for Lipid Metabolism and Atherosclerosis

About a century ago, Anitschkow performed the first atherosclerosis experiment using cholesterol-fed rabbits (Anitschkow 1983/1913). Now, rabbits are widely used for the studies of human hypercholesterolemia and atherosclerosis. This is because unlike mice and rats which have different lipoprotein metabolism system from humans, the lipid metabolism and lipoprotein features of rabbits resemble humans (Fan and Watanabe 2003). Rabbits are sensitive to cholesterol-rich diet and develop hypercholesterolemia rapidly and aortic atherosclerotic lesions which are pathologically similar to those of humans. In this respect, rabbits have been used for the study of the expression of leucocytes-adhesion molecules on arterial endothelial cell surface (Cybulsky and Gimbrone 1991), the transformation of macrophages into foam cells in vivo (Rosenfeld et al. 1987), and the expression of cytokines and enzymes such as matrix metalloproteinases in atherosclerotic lesions (Galis et al. 1995; Ylä-Herrttuala et al. 1991; Fleet et al. 1992; Rosenfeld et al. 1992). In addition to cholesterol-fed rabbits, several rabbit models derived from naturally mutant rabbits were reported for the study of human hypercholesterolemia, including Watanabe heritable hyperlipidemic (WHHL) rabbits (Watanabe 1980), myocardial infarction-prone WHHL (WHHLMI) rabbits (Shiomi et al. 2003), St. Thomas' mixed hyperlipidemic (SMHL) rabbits (Ardern et al. 1999) and postprandial hypertriglyceridemic (PHT) rabbits (Kawai et al. 2006). One of the striking features of WHHLMI rabbits is that they developed severe coronary atherosclerosis and myocardial infarction (Shiomi et al. 2003). WHHL and WHHLMI rabbits have been used for the development of hypolipidemic compounds and anti-atherosclerotic drugs as well as for studies of lipoprotein metabolism and atherogenesis.

7.3.1 WHHL Rabbits

In 1973, Dr. Watanabe found a male JW rabbit showing hypercholesterolemia and later established a WHHL colony by his name, Watanabe heritable hyperlipidemic rabbits, in 1979 (Watanabe 1980). On chow diet, plasma levels of total cholesterol of WHHL rabbits were up to 500–900 mg/dl, which are about 20 fold-higher than normal JW rabbits. HPLC and ultracentrifugation analysis revealed that there is a marked accumulation of LDL fractions accompanied by reduced HDL fractions in WHHL rabbits. Later, Yamamoto et al. demonstrated that WHHL rabbits have defect in LDL receptors which is characterized by the deletion of 12 base pairs of nucleotide in the LDL receptor gene (Yamamoto et al. 1986), which explained why there was marked accumulation of LDL in the plasma of WHHL rabbits (Tanzawa et al. 1980; Kita et al. 1981). In addition, WHHL rabbits, had tenfold higher levels of plasma triglycerides than normal rabbits. The CETP activity was high in the plasma (Son and Zilversmit 1986) and expression of apoB-editing enzyme was not detected in liver (Kozarsky et al. 1996). Therefore, lipoprotein metabolism of WHHL rabbits resembles that of human familial hypercholesterolemia patients. Because of these features, WHHL rabbits contributed to studies aiming at elucidating human lipoprotein metabolism, especially the LDL receptor pathway (Goldstein et al. 1983). In addition, WHHL rabbits have been used to screen the lipid-lowering effects of many compounds (Table 7.2). The best example of these studies examined in WHHL rabbits is statin (Watanabe et al. 1981; Tsujita 1986; Watanabe et al. 1988; Shiomi et al. 1995, 2005; Shiomi and Ito 1999a), which is one of the most effective hypocholesterolemic drugs prescribed right now for more than 30 million patients in the world each year.

WHHL rabbits have also contributed to studies of atherosclerosis and examination of anti-atherosclerotic effects of several compounds (Table 7.2). Due to the LDL receptor deficiency and hypercholesterolemia, atherosclerotic lesions developed spontaneously in WHHL rabbits under a normal standard chow. Therefore, WHHL rabbits can be used for the study of atherosclerosis initiation and the characterization of atherosclerotic pathology (Shiomi 2008). Atherosclerotic lesions of the aorta of WHHL rabbits were fatty streak at weaning, became macrophage-rich lesions at young age, established lesion (fibro-atheroma) at mature age, and complicated lesions (accumulation extracellular lipids, thin fibrous cap, calcification, and intra-

Table 7.2 Hypolipidemic and anti-atherosclerotic studies using WHHL or WHHLMI rabbits

Compounds	Hyperlipidemia	Aortic lesion	Coronary lesion
Statin	○	×, ○	○
Resin	○	○	n.d.
Statin + Resin	○	○	○
Squalene synthase inhibitor	○	○	○
MTP inhibitor	○	n.d.	n.d.
ACAT inhibitor	×, ○	×, ○	×, ○
Probucol	○	○	n.d.
M-CSF or GM-CSF	○	○	n.d.
ApoE	○	○	n.d.
Fibrate	×	n.d.	n.d.
Fish oil	×, ○	×, ○	n.d.
Thiazolidinedione	×	Δ	Δ
Ca^{2+} antagonist	×	×	×
β-blocker	×	×	×
ACE inhibitor	×	○	n.d.
A-II receptor antagonist	×	○	n.d.
Gene therapy	○	n.d.	n.d.

○, effective; ×, not effective; n.d., not determined

plaque hemorrhage) at old age. These histological features resemble many respects of human atherosclerotic lesions. Using WHHL rabbits, Watanabe et al. demonstrated that plasma cholesterol reduction by pravastatin, an inhibitor of cholesterol synthesis, can delay development of atherosclerosis (Watanabe et al. 1988).

7.3.2 *WHHLMI Rabbits*

Although atherosclerotic lesions developed in aorta of every WHHL rabbit, the incidence of coronary atherosclerosis was very low and myocardial infarction did not occur. To develop coronary atherosclerosis in WHHL rabbits, Watanabe et al. (Watanabe et al. 1985) started to develop coronary atherosclerosis-prone WHHL rabbits by selective breeding in 1995. Studies using the coronary atherosclerosis-prone WHHL rabbits demonstrated that reduction of plasma cholesterol levels by statins improved composition of coronary plaques from unstable lesions to stable lesions (Shiomi et al. 1995, 2005; Shiomi and Ito 1999a). However, the incidence of myocardial infarction was still very low due to macrophage-poor lesions. In quantitative analyses of atherosclerotic lesions using imaging techniques (Shiomi et al. 1994), the composition of coronary atherosclerosis differed from the aortic ones. Thereafter, Shiomi et al. (Shiomi et al. 2003) started selective breeding from 1993 using indices of macrophage-rich coronary lesions, severe coronary stenosis, higher plasma cholesterol levels, and development of myocardial infarction. Seven years after, the incidence and severity of coronary lesions increased. The cumulated incidence of myocardial infarction in 30 months old animals was increased from 23% of the original WHHL rabbits to 97% in WHHLMI rabbits by the selective breeding (Shiomi et al. 2003). The degree of coronary stenosis (cross-sectional narrowing) at the age of 10–14 months also increased from 38% in the original WHHL rabbits to 82% in the WHHLMI rabbits (Ito et al. 2004). Myocardial infarction of WHHLMI rabbits is classified into four types: subendocardial infarction, intramural infarction, transmural infarction, and subepicardial infarction. In WHHLMI rabbits, fresh myocardial lesions consisted of hyperemia, eosinophilic degeneration of myocardial cells and infiltration of inflammatory cells was observed in the vicinity of old myocardial lesions consisting of myocardial fibrosis, scar, and dissolution of myocardial cells. The electrocardiograms from a WHHLMI rabbit monitored immediately before sudden decease showed an elevation of ST-segment and deep Q-wave which are the typical changes of acute myocardial infarction in humans. The coronary lesions of WHHLMI rabbits show various types, vulnerable plaque (large lipid core covered with thin fibrous cap), macrophage-rich plaque, complicated plaque and fibrous plaque, all similar to human coronary plaques. However, rupture of coronary plaque and/or formation of thrombosis were not observed. The next theme is the development of acute coronary syndrome in WHHLMI rabbits, with the coronary plaque rupture and the generation of thrombosis.

Figure 7.1 summarizes the characteristics of WHHLMI rabbits and their implementation in translational researches. LDL accumulates in the plasma due to LDL receptor deficiency, the low HDL-cholesterol and the metabolic syndrome-like

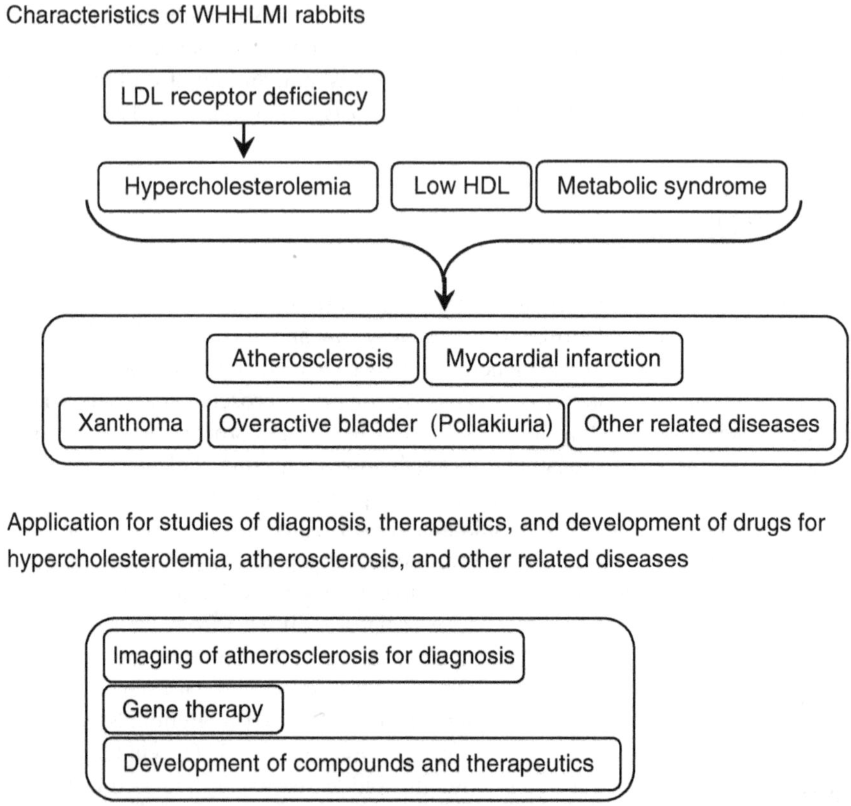

Fig. 7.1 Characteristics of WHHLMI rabbits and the application

feature (accumulation of visceral fats and hyperinsulinemia) which cause the development of atherosclerotic lesions in aorta, coronary arteries, pulmonary arteries, carotid arteries, cerebral arteries and other arteries. Myocardial infarction occurs due to coronary occlusion by atherosclerotic lesions. In addition, xanthoma, overactive bladder and other diseases appeared related to hypercholesterolemia and atherosclerosis.

Ideally, it is essential that animal models should contribute not only to the mechanistic study of human diseases but also to the development of new drugs, devices, or techniques for therapeutics. As mentioned above, several hypocholesterolemic drugs have been developed and the anti-atherosclerotic effects of some of them could be proved. Statin is an inhibitor of HMG-CoA reductase, a rate-limiting enzyme in the cholesterol biosynthesis pathway. Lipid-lowering effect of statin was initially detected in WHHL rabbits and it is the most potent drugs to prevent acute coronary syndromes (Naghavi et al. 2003). Studies using WHHL rabbits contributed to understand how the reduction of the serum cholesterol levels stabilized atherosclerotic lesions (Shiomi et al. 1995). Recently, WHHLMI rabbits have been used in studies based on imaging of atherosclerotic lesions by MRI (Meding et al. 2007; Steen et al. 2007), PET (Ogawa et al. 2006; Ishino et al. 2007) and intravascular ultrasound (IVUS) (Iwata et al. 2007). These techniques are promising for the identification of patients with coronary atherosclerosis to prevent acute coronary syndromes. In addition, WHHL (Kobe colony) and WHHLMI rabbits also showed metabolic syndrome-like metabolic disorders, including the accumulation of visceral fats (Shiomi et al. 1999b), insulin resistance (Zhang et al. 1991; Shiomi et al. 1999b) and mildly hyperglycedemia. Some WHHLMI rabbits also showed overactive bladder and keratopathy (Garibaldi and Goad 1988).

7.3.3 Other Rabbit Models for Human Lipid Disorders

In 1987, La Ville et al. (La Ville et al. 1987) developed a rabbit strain characterized by hypercholesterolemia (394 ± 100 mg/dl) with moderately increased or normal triglyceride levels. This rabbit strain showed elevated endogenous lipoproteins (very low-density lipoprotein (VLDL), intermediate density lipoprotein (IDL), and low-density lipoprotein (LDL)) cholesterol and LDL triglyceride levels due to over-secretion of VLDL from the liver. The LDL receptors functioned normally. Atherosclerosis developed in the aorta of these animals. These rabbits resemble human familial combined hyperlipidemia, which is one of the most common discrete hyperlipidemic disorders. Later on, this rabbit strain was bred into a new colony, the St. Thomas' mixed hyperlipidemic (SMHL) rabbit (Ardern et al. 1999). In SMHL rabbits, the serum cholesterol levels were 264 ± 68 mg/dl, and the triglyceride levels were 290 ± 55 mg/dl in feeding of very low cholesterol containing diet (0.08% cholesterol enriched diet) (De Roos et al. 2005). SMHL rabbits also showed insulin resistance (De Roos et al. 2001).

Ito and coworkers developed postprandial hypertriglyceridemic (PHT) rabbits (Kawai et al. 2006), which showed postprandial hypertriglyceridemia and metabolic syndrome-like feature. In PHT rabbits, the preprandial plasma triglyceride levels (4.55 ± 1.32 mmol/l) were about tenfold higher than in normal JW rabbits. After 15 h of continuously available feed of normal standard chow, the plasma triglyceride levels were markedly increased (15.9 ± 2.7 mmol/l) compared with the normal Japanese white rabbits (0.71 ± 0.01 mmol/l). In PHT rabbits, accumulation of visceral fat was prominent and insulin resistance was observed. These rabbits may also be important for the study of hyperlipidemia and metabolic syndrome.

7.4 Rabbit Models for Other Human Diseases

Rabbits also contributed to studies of cardiac function, infectious diseases, orthopedic surgery, cardiovascular surgery, and neoplastic diseases.

7.4.1 *Transgenic Rabbit Models for Prolonged QT and Sudden Cardiac Death in Human*

Recently, Brunner et al. (Brunner et al. 2008) reported that the expression of pore mutants of the human genes KCNQ1 and KCNH2 in rabbit hearts led to the generation of transgenic rabbits (LQT1 and LQT2) with a long QT phenotype. KCNQ1 and KCNH2 are genes encoding repolarizing K^+ channels. The human long-QT syndrome is characterized by delayed ventricular repolarization, prolonged QT-interval, ventricular arrhythmia and sudden cardiac death. Current murine transgenic animal models of long-QT syndrome are limited by substantial differences in cardiac electrophysiology. LQT2 rabbits showed a high incidence of spontaneous sudden cardiac death (>50% at 1 year) due to polymorphic ventricular tachycardia. Optical mapping revealed increased spatial dispersion of repolarization underlying arrhythmia. These transgenic rabbits have been used as a model to detect the channel-blocking properties of some anesthetic agents (Odening et al. 2008). The transgenic LQT1 and LQT2 rabbits are thus suitable animal models for studying the human prolonged QT syndrome.

7.4.2 *Rabbit Models for Thrombogenesis*

An important event after plaque rupture is thrombogenesis. Development of thrombus under physiological conditions is rare in animal models. Recently, Yamashita et al (Yamashita et al. 2004) reported that a blood flow reduction after thrombus propagation related to arterial thrombogenecity in rabbit model after single and repeated balloon injuries. In their study, after balloon injury of the normal femoral artery of normal rabbits, the blood flow was reduced. As a result, increased vascular wall thrombogenicity together with a substantial blood flow reduction are crucial for occlusive thrombus formation and the von Willebrand factor plays an important role in thrombus propagation. Finally, they speculated that reduced blood flow at plaque disruption sites might contribute to thrombus propagation leading to acute coronary syndromes. In coronary thrombi obtained by an aspiration device from patients with acute myocardial infarction, Hoshiba et al. (2006) observed co-localization of the von Willebrand factor with platelet thrombi, the tissue factor, platelets with fibrin and the consistent presence of inflammatory cells (CD16-positive neutrophils, CD45-positive mononuclear cells and CD34-positive blood progenitor cells). These studies suggest that, to propagate an arterial thrombus after a plaque rupture, a reduction of blood flow is important as well as the growing of the thrombus with inflammatory cells and erythrocytes. Therefore, rabbits are useful models for studies of thrombogenesis. Development of transgenic rabbits for factors of thrombogenesis may also play an important role.

7.4.3 Rabbit Models for Infectious Diseases and Deficiency of Immunological System

Manabe et al. (Manabe et al. 2008) showed that New Zealand white rabbits are useful animal for human latent tuberculosis. The global epidemic of tuberculosis claims more than two million lives yearly. *Mycobacterium tuberculosis* latently infects one third of the world population. They examined aerosol-infected rabbits with *Mycobacterium tuberculosis* and showed the formation of caseous lung granulomas which are strikingly similar to tuberculous lung lesions in humans. The lung burden of infection peaked at 5 weeks after aerosol infection followed by a host containment of infection that occurred in all rabbits. Corticosteroid-induced immunosuppression initiated after the disease containment resulted in a reactivation of disease. They also characterized the lung cellular immune response to inhaled *Mycobacterium tuberculosis* in the susceptible inbred Thorbecke rabbit (the genomically sequenced strain) and compared it to outbred, *Mycobacterium tuberculosis*-resistant, New Zealand white rabbits (Mendez et al. 2008). The development and severity of the immune reconstitution inflammatory syndrome was dependent on the antigen load at the time of immunosuppression and the subsequent bacillary replication during the corticosteroid-induced immunosuppression. This corticosteroid model is the only animal model to study the immune reconstitution inflammatory syndrome. The lung granulomas of inbred rabbits had a significantly higher number of cells expressing MHC Class II and CD11b, and a lower number of CD8 + T cells than the outbred controls. Effective utilization of this rabbit model could lead to a new tuberculosis diagnostic as well as to the elucidation of important correlates of protective immunity.

Human papillomavirus infections result in more than 250,000 deaths from cervical cancer in women worldwide. Hu et al. (2007) established a rabbit transgenic model expressing the human major histocompatibility complex (MHC-I) gene (HLA-A2.1). These transgenic rabbits expressed the HLA protein at a high level and HLA-A2.1 restricted rabbit CD8 cells were induced in these animals. Southern blot analysis demonstrated that the HLA-A2.1 gene was integrated into the rabbit genome and similar expression patterns of HLA-A2.1 and rabbit MHC class-I was observed in the three lines of transgenic rabbits. They demonstrated that HLA-A2.1 transgenic rabbits showed a susceptibility to cottontail rabbit papillomavirus infection akin to that of normal domestic rabbits. They also reported that a human papillomavirus type 16 E7 epitope can be engineered to be introduced into the cottontail rabbit papillomavirus E7 gene of the rabbit papillomavirus genome. This hybrid genome retained the ability to initiate paplillomas. The cottontail rabbit papillomavirus/HLA-A2.1 rabbit model has the potential to be used to screen HLA-A2.1-restricted immunogenic epitopes from human papillomaviruses in the context of in vivo papillomavirus infection. These studies suggested that rabbit is an excellent model to assess both natural and induced immunity to papillomavirus infections and that the transgenic rabbits may have utility for assessment of immunity to other human pathogens that are permissive in rabbits.

Rother reported the existence of a mutant rabbit which was deficient in the sixth component of complement (Rother et al. 1966). Current studies based on the use of the 6 complement-deficient rabbits suggested that this component is involved in the activity of the immune system, the activation of the inflammatory response and the hemolytic activity. Chartrand et al. (1979) demonstrated that delayed rejection was observed in the puppy hearts engrafted to C6-deficient rabbits. Schmiedt et al. (1998) suggested that C6-deficient rabbits delayed the development of atherosclerosis by cholesterol feeding due to the weak inflammatory responses of arterial cells. Therefore, C6-deficient rabbits may contribute to study human diseases related to the immune system and the inflammatory responses.

7.4.4 Rabbit Models for Human Articular Lesions and Therapeutics

Patients with articular cartilage lesions caused by injury or degenerative joint diseases become increased recently and these defects do not repair spontaneously. Studies using rabbits have been contributed to tentatively develop therapeutics against the diseases. Ikeda et al. (2009) showed that rabbits are useful to examine the effect on porosity and of the mechanical properties of a synthetic polymer (DL-lactide-coglycolide) scaffold on repair of osteochondral defects. They treated rabbits suffering from osteochondral defects in the femoral condyle with three types of scaffolds. Their study suggested that higher porosity allowing bone marrow cells to migrate to the scaffold is important in repairing osteochondral defects. Nakayama et al. (2009) performed a mechanical analysis of the effects of fibroblast growth factor-2 (FGF-2) on autologous osteochondral transplantation in an artificial rabbit model. They induced a full-thickness cartilage defect in the right femoral condyle and treated with osteochondral transplantation using an osteochondral plug taken from the left femoral condyle. Autologous osteochondral grafts transplanted with gelatin hydrogel containing FGF-2 acquired adequate stiffness at early postoperative phase. In addition, Ishida et al. (2007) demonstrated that platelet-rich plasma enhances the healing of meniscal defects in rabbits. These studies demonstrate that rabbits are useful for studies to develop therapeutics about human articular lesions.

7.4.5 Rabbit Models for Vascular Surgery

Rabbits are also useful for studies to develop technique for vascular and respiratory surgery. Kawanishi et al. (2007) showed that the prevention of back-bleeding from intercostal arteries and lumbar arteries during thoracoabdominal aortic surgery was considered to reduce spinal ischemic injury. They examined the effects of back-bleeding in spinal cord by comparing rabbits without back-bleeding from the lumbar arteries by draining from the aorta during aortic clamping with rabbits in which back-bleeding was not drained. Forty-eight hours later, the number of TUNEL-positive cells in rabbits draining back-bleeding was significantly smaller than those in rabbits with back-bleeding. Hasegawa et al. (2007) implanted autologous fibrin-coated vascular prostheses and/or xenologous fibrin-coated vascular prostheses in the bilateral carotid arteries of JW rabbits. As a result, autologous fibrin coating in thrombin-free fibrin-coated vascular prostheses improved antithrombogenecity. Their study suggested that autologous fibrin coating in thrombin-free fibrin-coated vascular prostheses have a potential for clinical use in hybrid small-caliber vascular grafts.

7.4.6 Rabbit Models for Tumor Study

Several transplantable rabbit tumors have been reported and VX2 tumor has been used in some studies. The VX2 carcinoma arose as the result of spontaneous transformation of a virus-induced skin papilloma in a domestic rabbit (Kidd and Rous 1940) and is a type of dermatological squamous cancer induced by the Shope virus. In general, the VX-2 tumor had a high malignant potential, with capacity for rapid reproduction, infiltration and metastasis. VX2 tumors have been transplanted into lung, liver, and other organs in rabbits. These rabbit VX2 tumors have been studied for establishing therapeutics and diagnostics. Virmani et al. (2008) demonstrated that hypoxia caused by transcatheter arterial embolization of VX2 liver tumors activates the hypoxia-inducible factor-1 alpha, a transcription factor that in turn regulates other pro-angiogenic factors. Jiang et al. (2008) reported that the hemodynamic changes in the liver caused by rabbit VX2 liver tumor can be detectable after tumor inoculation and that functional CT can evaluate the physiological characteristics of early angiogenesis. In addition, Ohira et al. (2008) demonstrated that FDG-PET was useful for monitoring the early effects of radiofrequency ablation in VX2 rabbit tumors implanted into the back muscles. These studies have demonstrated that rabbits are also useful for studies of tumor.

7.5 Conclusions

Genetically modified mice is playing an essential role in the clarification of the expression and functions of individual gene. However, to translate or extrapolate the results of animal studies to humans, it is necessary that key gene expression and function are equivalent to human rather than the outward features or phenotype. For example, to study human hypercholesterolemia, not only the hyperlipidemia itself, the characteristics of the lipoprotein profiles and enzymes in the lipoprotein metabolism of animal models is vitally important for translational researches. In addition, histopathological and/or immunohistochemical features similar to humans are also important. Several rabbit models for some human diseases described in this chapter are useful in translational researches. Applying these animal models in translational researches will contribute to elucidation of the mechanism of human diseases and development of novel compounds, therapeutics, or diagnostic instruments containing lesion imaging techniques.

Acknowledgements This work was supported partly by a research grant from the Health and Labor Ministry of Japan and unrestricted research grants from Sankyo Co. Ltd., Japan.

References

Anitschkow NN, Chalatow S (Translated by Pelias MZ). (1983) On experimental cholesterin steatosis and its significance in the origin of some pathological processes. Arteriosclerosis 3: 178–182. (The original article: Ueber experimentelle cholesterinsteatose und ihre bedeutung für die entstehung einiger pathologischer prozesse. Zentralbl Allg Pathol Pathol Anat 1913; 24: 1–9.)

Ardern HA, Benson GM, Suckling KE, Caslake MJ, Shepherd J, Packard CJ. (1999) Apolopoprotein B overproduction by the perfused liver of the St. Thomas' mixed hyperlipidemic (SMHL) rabbit. J Lipid Res 40: 2234–2243.

Brunner M, Peng X, Liu GX, Ren Z, Ziv O, Choi B, Mathur R, Hajjiri M, Odening KE, Steinberg E, Folco EJ, Pringa E, Centracchio J, Macharzina RR, Donahay T, Schofield L, Rana N, Kirk M, Mitchell GF, Poppas A, Zehender M, Koren G. (2008) Mechanisms of cardiac arrhythmias and sudden death in transgenic rabbits with long QT syndrome. J Clin Invest 118: 2246–2259.

Chartrand C, O'Regan S, Robitaille P, Pino-Blonde M. (1979) Delayed rejection of cardiac xenografts in C6-defficient rabbits. Immunology 38: 245–248.

Cybulsky MI, Gimbrone MA. (1991) Endothelial expression of a mononuclear leukocyte adhesion molecule during atherogenesis. Science 251: 788–791.

De Roos B, Caslake MJ, Ardern HA, Benson GM, Suckling KE, Packard CJ. (2001) Insulin resistance in the St. Thomas' mixed hyperlipidemic (SMHL) rabbit, a model for familial combined hyperlipidemia. Atherosclerosis156: 249–254.

De Roos B, Caslake MJ, Milliner K, Benson GM, Suckling KE, Packard CJ. (2005) Characterisation of the lipoprotein structure in the St. Thomas' mixed hyperlipidemic (SMHL) rabbit. Atherosclerosis 181: 63–68.

Fan J, Watanabe T. (2003) Transgenic rabbits as therapeutic protein bioreactors and human disease models. Pharmacol Ther 99: 261–282.

Fleet JC, Clinton SK, Salomon RN, Loppnow H, Libby P. (1992) Atherogenic diets enhance endoxin-stimulated interleukin-1 and tumor necrosis factor gene expression in rabbit aorta. J Nutr 122: 294–305.

Galis ZS, Sukhova GK, Kränzhofer R, Clark S, Libby P. (1995) Macrophage foam cells from experimental atheroma constitutively produce matrix-degradation proteins. Proc Natl Acad Sci USA 92: 402–406.

Garibaldi BA, Goad MEP. (1998) Lipid keratopathy in the Watanabe (WHHL) rabbit. Vet Pathol 25: 173–174.

Goldstein JL, Kita T, Brown MS. (1983) Defective lipoprotein receptors and atherosclerosis: lessons from an animal counterpart of familial hypercholesterolemia. N Engl J Med 309: 288–296.

Hasegawa T, Okada K, Takano Y, Hiraishi Y, Okita Y. (2007) Autologous fibrin-coated small-caliber vascular prostheses improve antithrombogenecity by reducing immunologic response. J Thorac Cardiovasc Surg 133: 1268–1276.

Hoshiba Y, Hatakeyama K, Tanabe T, Asada Y, Goto S. (2006) Co-localization of von Willebrand factor with platelet thrombi, tissue factor and platelets with fibrin, and consistent presence of inflammatory cells in coronary thrombi obtained by an aspiration device from patients with acute myocardial infarction. J Thromb Haemost 3: 1140–1120.

Hu J, Peng X, Budgeon LR, Cladel NM, Balogh KK, Christensen ND. (2007) Establishment of a cottontail rabbit papillomavirus/HLA-A2.1 transgenic rabbit model. J Virol 81: 7171–7177.

Ikeda R, Fujioka H, Nagura I, Kokubu T, Toyokawa N, Inui A, Makino T, Kaneko H, Doita M, Kurosaka M. (2009) The effects of porosity and mechanical property of a synthetic polymer scaffold on repair of osteochondral defects. Int Orthop (in press).

Ishida K, Kuroda R, Miwa M, Tabata Y, Hokugo A, Kawamoto T, Sasaki K, Doita M, Kurosaka M. (2007) The regenerative effects of platelet-rich plasma on meniscal cells in vitro and its in vivo application with biodegradable gelatin hydrogel. Tissue Eng 13: 1103–1112.

Ishino S, Mukai T, Kume N, Asano D, Ogawa M, Kuge Y, Minami M, Kita T, Shiomi M, Saji H. (2007) Lectin-like oxidized LDL receptor-1 (LOX-1) expression is associated with atherosclerotic plaque instability-analysis in hypercholesterolemic rabbits. Atherosclerosis 195: 48–56.

Ito T, Yamada S, Shiomi M. (2004) Progression of coronary atherosclerosis relates to the onset of myocardial infarction in an animal model of spontaneous myocardial infarction (WHHLMI rabbits). Exp Anim 53: 339–334.

Iwata A, Miura S, Imaizumi B, Saku K. (2007) Measurement of atherosclerotic plaque volume in hyperlipidemic rabbit aorta by intravascular ultrasound. J Cardiol 50: 229–234.

Jiang HJ, Zhang ZR, Shen BZ, Wan Y, Guo H, Shu SJ. (2008) Functional CT for assessment of early vascular physiology in liver tumors. Hepatobiliary Pancreat Dis Int 7: 497–502.

Kawai T, Ito T, Ohwada K, Mera Y, Matsushita M, Tomoike H. (2006) Hereditary postprandial hypertriglyceridemic rabbit exhibits insulin resistance and central obesity: a novel model of metabolic syndrome. Arterioscler Thromb Vasc Biol 26: 2752–2757.

Kawanishi Y, Okada K, Tanala H, Yamashota T, Nakagiri K, Okita Y. (2007) The adverse effect of back-bleeding from lumber arteries on spinal cord pathophysiology in a rabbit model. J Thorac Cardiovasc Surg 133: 1553–1558.

Kidd JG, Rous P. (1940) A transplantable rabbit carcinoma originating in a virus-induced papilloma and containing the virus in masked or altered form. J Exp Med 71: 813–838.

Kita T, Brown MS, Watanabe Y, Goldstein JL. (1981) Deficiency of low density lipoprotein receptors in liver and adrenal gland of the WHHL rabbit, an animal model of familial hypercholesterolemia. Proc Natl Acad Sci USA 78: 2268–2272.

Kozarsky KF, Bonen DK, Giannoni F, Funahashi T, Wilson JM, Davivson NO. (1996) Hepatic expression of the catalytic subunit of the apolipoprotein B mRNA editing enzyme (apobec-1) ameliorates hypercholesterolemia in LDL receptor-deficient rabbits. Hum Gene Ther 7: 973–977.

La Ville A, Turner PR, Pittilo RM, Martini S, Marenah CB, Rowles PM, Morris G. (1987) Thomson GA, Woolf N, Lewis B. Hereditary hyperlipidemia in the rabbit due to overproduction of lipoproteins. I. Biochemical studies. Arteriosclerosis 7: 105–112

Manabe YC, Kesavan AK, Lopez-Molina J, Hatem CL, Brooks M, Fujiwara R, Hochstein K, Pitt MLM, Tufariello J, Chan J, McMurray DN, Bishai WR, Dannenberg AM, Mendez S. (2008) The aerosol rabbit model of TB latency, reactivation and immune reconstitution inflammatory syndrome. Tuberculosis 88: 187–196.

Mendez S, Hatem CL, Kesavan AK, Lopez-Molina J, Pitt ML, Dannenberg AM Jr, Manabe YC. (2008) Susceptibility to tuberculosis: composition of tuberculous granulomas in Thorbecke and outbred New Zealand white rabbits. Vet Immunol Immunopathol 122: 167–174.

Meding J, Urich M, Licha K, Reinhardt M, Misselwitz B, Fayad ZA, Weinmann HJ. (2007) Magnetic resonance imaging of atherosclerosis by targeting extracellular matrix deposition with Gadofluorine M. Contrast Media Mol Imag 2: 120–129.

Naghavi M, Libby P, Falk E, Casscells SW, Litovsky S, Rumberger J, Badimon JJ, Stefanadis C, Moreno P, Pasterkamp G, Fayad Z, Stone PH, Waxman S, Raggi P, Madjid M, Zarrabi A, Burke A, Yuan C, Fitzgerald PJ, Siscovick DS, de Korte CL, Aikawa M, Airaksinen KEJ, Assmann G, Becker CR, Chesebro JH, Farb A, Galis ZS, Jackson C, Jang I, Koenig W,

Lodder RA, March K, Demirovic J, Navab M, Priori SG, Rekhter MD, Bahr R, Grundy SM, Mehran R, Colombo A, Boerwinkle E, Ballantyne C, Insull W Jr, Schwartz RS, Vogel R, Serruys PW, Hansson GK, Faxon DP, Kaul S, Drexler H, Greenland P, Muller JE, Virmani R, Ridker PM, Zipes DP, Shah PK, and Willerson JT. (2003) From vulnerable plaques to vulnerable patients, a call for new definitions and risk assessment strategies: Part I. Circulation 108: 1664–1672.

Nakayama J, Fujioka H, Nagura I, Kokubu T, Makino T, Kuroda R, Tabata Y, Kurosaka M. (2009) The effect of fibroblast growth factor-2 on autologous osteochondral transplantation. Int Orthop 33: 275–280.

Odening KE, Hyder O, Chaves L, Schofield L, Brunner M, Kirk M, Zehender M, Peng X, Koren G. (2008) Pharmacogenomics of anesthetic drugs in transgenic LQT1 and LQT2 rabbits reveal genotype-specific differential effects on cardiac repolarization. Am J physiol Heart Circ Physiol 295: H2264–H2272.

Ogawa M, Magata Y, Kato T, Hatano K, Ishino S, Mukai T, Shiomi M, Ito K, Saji H. (2006) Application of 18F-FDG PET for monitoring the therapeutic effect of antiinflammatory drugs on stabilization of vulnerable atherosclerotic plaques. J Nucl Med 47: 1845–1850.

Ohira T, Okuma T, Matsuoka T, Wada Y, Nakamura K, Watanabe Y, Inoue Y. (2009) FDG-MicroPET and diffusion-weighted MR image evaluation of early changes after radiofrequency ablation in implanted VX2 tumors in rabbits. Cardiovasc Intervent Radiol 32: 114–120.

Rosenfeld ME, Tsukada T, Gown AM, Ross R. (1987) Fatty streak initiation in Watanabe heritable hyperlipidemic and comparably hypercholesterolemic fat-fed rabbits. Arteriosclerosis 7: 9–23.

Rosenfeld ME, Ylä-herttuala S, Lipton BA, Ord VA, Witztum JL, Steinberg D. (1992) Macrophage colony-stimulating factor mRNA and protein in atherosclerotic lesions of rabbits and humans. Am J Pathol 140: 291–300.

Rother K, Rother U, Muller-Eberhard HJ, Nilsson UR. (1966) Defficiency of the sixth component of complement in rabbits with an inherited complement defect. J Exp Med 124: 773–785.

Schmiedt W, Kinscherf R, Deigner HP, Kamencic H, Nauen O, Kilo J, Oelert H, Metz J, Bhakdi S. (1998) Complement C6 deficiency protects against diet-induced atherosclerosis in rabbits. Arterioscler Throm Vasc Biol 18: 1790–1795.

Shiomi M, Fan J. (2008) Unstable coronary plaques and cardiac events in myocardial infarction-prone Watanabe heritable hyperlipidemic (WHHLMI) rabbits: questions and quandaries. Curr Opin Lipidol 19: 631–636.

Shiomi M, Ito T. (1999a) Effect of cerivastatin sodium, a new inhibitor of HMG-CoA reductase, on plasma lipid levels, progression of atherosclerosis, and the lesional composition in the plaques of WHHL rabbits. Br J Pharmacol 126: 961–968.

Shiomi M, Ito T, Tsukada T, Tsujita Y, Horikoshi H. (1999b) Combination treatment with troglitazone, an insulin action enhancer, and pravastatin, an inhibitor of HMG-CoA reductase, shows synergistic effect on atherosclerosis of WHHL rabbits. Atherosclerosis 42: 345–353.

Shiomi M, Ito T, Tsukada T, Yata T, Ueda M. (1994) Cell compositions of coronary and aortic atherosclerotic lesions differ; An immunohistochemical study. Arterioscler Thromb 14: 931–937.

Shiomi M, Ito T, Tsukada T, Yata T, Watanabe Y, Tsujita Y, Fukami M, Fukushige J, Hosokawa T, Tamura A. (1995) Reduction of serum cholesterol levels alters lesional composition of atherosclerotic plaques: effect of pravastatin sodium on atherosclerosis in mature WHHL rabbits. Arterioscler Thromb Vasc Biol 15: 1938–1944.

Shiomi M, Ito T, Yamada S, Kawashima S, Fan J. (2003) Development of an animal model for spontaneous myocardial infarction (WHHLMI Rabbit). Arterioscler Thromb Vasc Biol 23: 1239–1244.

Shiomi M, Yamada S, Ito T. (2005) Atheroma stabilizing effects of simvastatin due to depression of macrophages or lipid accumulation in the atheromatous plaques of coronary atherosclerosis-prone WHHL rabbits. Atherosclerosis 178: 287–294.

Son YC, Zilversmit DB. (1986) Increased lipid transfer activities in hyperlipidemic rabbit plasma. Arteriosclerosis 6: 345–351.

Steen H, Lima JA, Chatterjee S, Kolmakova A, Gao F, Rodrigues ER, Stuber M. (2007) High-resolution three-dimensional aortic magnetic resonance angiography and quantitative vessel wall characterization of different atherosclerotic stages in a rabbit model. Invest Radiol 42: 614–621.

Tanzawa K, Shimada Y, Kuroda M, Tsujita Y, Arai M, Watanabe Y. (1980) WHHL-rabbit: a low density lipoprotein receptor-deficient animal model for familial hypercholesterolemia. FEBS Lett 118: 81–84.

Tsujita Y, Kuroda M, Shimada Y, Tanzawa K, Arai M, Kaneko I, Tanaka M, Masuda H, Tarumi C, Watanabe Y, Fujii S. (1986) CS-514, a competitive inhibitor of 3-hydroxy-3-methylglutaryl coenzyme A reductase: tissue-selective inhibition of sterol synthesis and hypolipidemic effect on various animal species. Biochim Biophys Acta 877: 50–60.

Virmani S, Rhee TK, Ryu RK, Sato KT, Lewandowski RJ, Mulcahy MF, Kulik LM, Szolc-Kowalska B, Woloschak GE, Yang GY, Salem R, Larsin AC, Omary RA. (2008) Comparison of hypoxia-inducible factor-1alpha expression before and after transcatheter arterial ambolisation in rabbit VX2 liver tumors. J Vasc Interv Radiolo 19: 1483–1489.

Watanabe Y. (1980) Serial inbreeding of rabbits with hereditary hyperlipidemia (WHHL-rabbit). Atherosclerosis 36: 261–268.

Watanabe Y, Ito T, Saeki M, Kuroda M, Tanzawa K, Mochizuki M, Tsujita Y, Arai M. (1981) Hypolipidemic effects of CS-500 (ML-236B) in WHHL-rabbit, a heritable animal model for hyperlipidemia. Atherosclerosis 38: 27–31.

Watanabe Y, Ito T, Shiomi M. (1985) The effect of selective breeding on the development of coronary atherosclerosis. Atherosclerosis 56: 71–79.

Watanabe Y, Ito T, Shiomi M, Tsujita Y, Kuroda M, Arai M, Fukami M, Tamura A. (1988) Preventive effect of pravastatin sodium, a potent inhibitor of 3-hydroxy-3-methylglutaryl coenzyme A reductase, on coronary atherosclerosis and xanthoma in WHHL rabbits. Biochim Biophys Acta 960: 294–302.

Yamamoto T, Bishop RW, Brown MS, Goldstein JL, Russell DW. (1986) Deletion in cysteine-rich region of LDL receptor impedes transport to cell surface in WHHL rabbit. Science 232: 1230–1237.

Yamashita A, Furukoli E, Marutsuka K, Hatakeyama K, Yamamoto H, Tamura S, Ikeda Y, Sumiyoshi A, Asada Y. (2004) Increased vascular wall thrombogenicity combined with reduced blood flow promotes occlusive thrombus formation in rabbit femoral artery. Arterioscler Thromb Vasc Biol 24: 2420–2424.

Ylä-Herrttuala S, Lipton BA, Rosenfeld ME, Särkioja T, Yoshimura T, Leonard EJ, Witztum JL, Steiberg D. (1991) Expression of monocyte chemoattractant protein 1 in macrophage-rich areas of human and rabbit atherosclerotic lesions. Proc Natl Acd Sci USA 88: 5252–5256.

Zhang B, Saku K, Hirata K, Liu R. (1991) Tateishi K, Yamamoto K, Arakawa K. Insulin resistance observed in WHHL rabbits. Atherosclerosis 91: 277–278.

Chapter 8
Transgenic Rabbits to Prepare Pharmaceutical Proteins

Louis-Marie Houdebine, Geneviève Jolivet, and Pierre-Jean Ripoll

Abstract The preparation of recombinant pharmaceutical proteins is one of the major challenges of biotechnology. Mammalian cells are required for a number of proteins which must be modified posttranscriptionally. Animal cell lines cultured in fermentors are presently the major source of complex proteins. The milk of transgenic animals proved to be a possible source of pharmaceutical proteins and one of them, human antithrombin III, has been approved by the EU (EMEA) and US (FDA) medicament agencies. Several species are being implemented for this purpose. Rabbits are one of these species. It offers several advantages: low cost to produce transgenic founders, rapid reproduction, easy and cheap scaling up, easy breeding in pathogen-free conditions and insensitivity to prion diseases. Rabbits are thus an efficient tool to prepare several kilograms of a recombinant protein per year.

Keywords Pharmaceutical proteins, transgenic, blood, milk, antibodies

8.1 Introduction

The production of recombinant pharmaceutical proteins is one of the successes of biotechnology. Most of these proteins are presently prepared in bacteria or in mammalian cells such as CHO or BHK. Other systems may produce recombinant proteins (Houdebine 2009). Among these systems are yeast, insect cells infected by recombinant baculovirus, transgenic plants and transgenic animals. Transgenic animals or plants are particularly attractive as they contain a very high number of different cell types maintained in ideal physiological conditions. Animals of different size may be used. Different transgenic species and organs are thus being

L.-M. Houdebine (✉) and G. Jolivet
Biologie du Développement et Reproduction, Institut National de la Recherche Agronomique, 78350 Jouy en Josas, France
e-mail: louis.houdebine@jouy.inra.fr

P.-J. Ripoll
BioProtein Technologies Domaine de Vilvert 78350 Jouy en Josas, France

L.-M. Houdebine and J. Fan (eds.), *Rabbit Biotechnology: Rabbit Genomics, Transgenesis, Cloning and Models*,

exploited or under study. The most promising systems are transgenic mammals as they are expected to be the most appropriate to proceed to the posttranslational modifications of proteins, such as glycosylations. The most likely mammalian source of recombinant proteins are milk and to a much lower degree blood. Recent studies have shown that egg white from transgenic chicken may be a good source of recombinant proteins (Van de Lavoir et al. 2006; Lillico et al. 2007; Han 2008).

The first time transgenic animals were considered as a potential source of recombinant proteins was in 1982 (Palmiter et al. 1982) after the generation of giant transgenic mice over expressing rat or human growth hormone genes. The hormone was found in the blood of some mice at relatively high concentration, up to 50 µg/ml. This suggested that blood of transgenic animals could be a very abundant source of pharmaceutical proteins.

The first transgenic rabbits were obtained in 1985 (Hammer et al. 1985) as larger animals, sheep and pigs. This data revealed that the DNA microinjection method could be used to generate various transgenic mammals and not only mice. This reinforced the idea that transgenic animals, namely of farm animals, could be the source of pharmaceutical proteins. The hypothesis that milk could be a better system than blood to obtain large amounts of recombinant proteins became particularly attractive and the proof of concept was given in 1987 with the demonstration that biologically active human plasminogen activator (tPA) could be produced in the milk of transgenic mice (Gordon et al. 1987). The different transgenic mammals considered as able to produce large amounts of recombinant proteins in their milk are cows, sheep, goats, pigs and rabbits. Indeed, the success of the method is dependent on several factors: the efficiency of transgenesis and the capacity to secrete high amounts of well-matured recombinant proteins. (Van de Lavior et al. 2006; Lillico et al. 2007; Han 2008).

This chapter reports the different projects which have undertaken for using transgenic rabbits to produce pharmaceutical proteins.

8.2 Techniques to Generate Transgenic Rabbits

DNA microinjection was the first technique used to generate transgenic rabbits (Hammer et al. 1985). It remains the only being used in practice. This is due to the fact that rabbit embryos are abundant, as easily microinjected and transferred to foster mothers as mouse embryos. The yield of transgenesis is significantly lower than in mice but it corresponds to 1% of the microinjected embryos. Microinjection is thus not too laborious and this allowed a few groups to generate transgenic rabbits to create models for studying human diseases and to prepare recombinant proteins (Bösze and Houdebine 2006).

This relatively high efficiency to generate transgenic rabbits even with long genomic DNA fragments dissuaded researchers to use other techniques available for other species (Houdebine 2008). Transposons which are well-appreciated tools to enhance gene integration rate have not been used so far in rabbits. Lentiviral vectors which proved highly efficient in various vertebrates are currently being tested in several laboratories.

More than a decade ago, it was shown that sperm incubated in the presence of DNA before being used for fertilization was able to transfer the foreign gene into the oocyte and generate transgenic mice. This method appeared difficult to use due to a frequent degradation of DNA (Smith and Spadafora 2005). Transgenic mice and rabbits were obtained by incubating sperm with DNA in the presence of DMSO (dimethylsuphoxide) and by using conventional in vitro fertilization (Shen et al. 2006). The method has been greatly improved, mainly by using ICSI (Intracytoplasmic Sperm Injection). This technique which consists of injecting sperm into the cytoplasm of oocytes is currently used for in vitro fertilization in humans. In order to transfer genes, sperm in which plasma membrane has been damaged by freezing and thawing were incubated in the presence of the gene of interest and further used for fertilization by ICSI. This method proved efficient in mice (Moreira et al. 2007; Shinohara et al. 2007) and pigs (Yong et al. 2006). Transposon use and ICSI may be combined to increase the yield of transgenesis (Shinohara et al. 2007). ICSI is therefore an excellent method to generate transgenic animals, on condition ICSI is possible in the considered species. One advantage of ICSI is that long fragments of DNA may be used to transfer the gene of interest. Another advantage is that foreign DNA is integrated at the fist cell stage of embryos. This reduces the number of animals being mosaic for the transgene. The use of ICSI to generate transgenic rabbits is currently under study.

Pluripotent cells known as ES (embryonic stem) cells have not yet been fully characterized and controlled to be currently used for gene transfer and particularly gene replacement by homologous recombination in rabbits (this issue).

Cloning by nuclear transfer proved to be possible in rabbits (Chesné et al. 2002). This method remains laborious (this issue) and it is not currently used for gene addition or replacement.

The recent obtention of mouse and human pluripotent cells (iPS) from somatic cells by transferring only three of the genes involved in the maintenance of natural pluripotent cells offers unprecedented possibility to establish and use pluripotent cell lines in rabbits (Pera and Hasegawa 2008).

8.3 Production in Blood

Serum, which collects secretion from many tissues, may be the source of recombinant proteins. Human α1-antitrypsin synthesized essentially in liver was thus obtained at a level as high as 1 mg/ml from the serum of transgenic rabbits. This protein seemed matured in an appropriate manner (Massoud et al. 1991). One limitation in this case was the difficulty to separate the recombinant from the endogenous protein. The replacement of the endogenous gene by the human genes using homologous recombination would solve this problem. The success obtained with α1-antitrypsin cannot be easily extended to all the proteins. Indeed, α1-antitrypsin is highly glycosylated and thus stable in blood at a high level. The same is not true for growth hormone. On the other hand, a human protein secreted in the blood is likely to act on a number of tissues and alter health of the transgenic animals.

Recombinant antibodies were also found in the blood of transgenic rabbits (Weidle et al. 1991; Limonta et al. 1995). However, the antibodies were present at a relatively low concentration and they were hybrid molecules containing chains from the endogenous antibodies.

Replacement of loci harbouring the antibody genes by human loci may allow animals to synthesize human antibodies. This has been achieved in mice to generate human monoclonal antibodies. In practice, the genes coding for a human monoclonal antibody can be isolated from the lymphocytes of immunized mice, engineered and used to produce the antibody at an industrial scale, using one of the available production systems. Another possibility is to replace immunoglobulin loci of an animal by the corresponding human loci. These animals can be immunized and used as an abundant source of polyclonal antibodies directed against the antigens. These polyclonal antibodies are expected to be potent tools to neutralize pathogens or induce tumour regression in patients. This ambitious project is being developed in cows, pigs and rabbits (Buelow and Schooten 2006).

8.4 Production in Milk

The first recombinant protein prepared in rabbit milk was human interleukin-2 (Buhler et al. 1990). This was achieved with αs1-casein promoter and the interleukin cDNA. The production level was low. A number of studies made it possible the improvement of the expression vectors. The promoters used in rabbits to produce recombinant proteins in milk are essentially those from ruminant αs1-and β-casein genes as well as from mouse and rabbit WAP (whey acidic protein) genes and sheep β-lactoglobulin. These promoters are strong and they proved to direct the expression of foreign cDNAs with variable efficiency. The casein promoters are greatly stimulated by enhancers present in the first intron of these genes. The rabbit WAP promoter used contained 6.3 kb and appeared more efficient than the mouse WAP promoter which contained only 2.6 kb. It was shown later that the rabbit 6.3 kb contained a potent enhancer 5.5 kb upstream of the cap site (Millot et al. 2001). A similar enhancer was found later in the 4.3 kb preceding the cap site of the mouse WAP gene.

The level of the proteins produced in milk using WAP promoters proved to be poorly predictable and dependent on the associated genes, as though the genes in question brought enhancers on a case by case basis. The addition of two copies in tandem of the 5'HS4 insulator from the chicken β-globin locus reduced markedly the intensity of the position effect on the transgene (Rival-Gervier et al. 2003; Giraldo et al. 2003). Yet, the expression of the transgenes driven by the rabbit WAP promoter remained too low and not constant.

Addition of various enhancers and of appropriate introns, mRNA stabilizers and transcription terminators in the vectors containing the rabbit WAP promoters (Houdebine 2007) and the 5'HS4 insulator improved greatly the expression of cDNAs in rabbit milk (Soler et al. 2005; Ripoll PJ et al. 2008). Up to several milligrams per millilitres of foreign proteins could be found in the different transgenic rabbit lines.

A study of the genomic DNA surrounding the WAP gene revealed that the fragment of 30 kb containing the WAP gene was very poorly capable of directing expression of the gene. On the contrary, longer fragments, up to 150 kb, were much more active and several regions of this genomic fragment containing potent enhancers were identified (Rival et al. 2001; Rival-Gervier et al. 2002; Saidi et al. 2007). Interestingly, expression of the reporter firefly luciferase cDNA was increased five- to tenfold when it was integrated into the 150 kb DNA fragment (Soler E et al.). A new family of efficient vectors can thus be derived from the BACs (bacterial artificial chromosome) containing the WAP gene.

The different promoters used to prepare pharmaceutical proteins in milk show a weak expression of the transgenes in tissues other than mammary gland and at variable levels according to the rabbit lines. This leaky expression has sometimes been observed also in males. This may be the source of deleterious effect for the animals. Human EPO (erythropoietin) is known to exert its biological effect at low concentration. The leaky expression of the EPO gene under the action of rabbit WAP promoter induced severe health problems in the rabbits (Massoud et al. 1996). In this particular case, the project was compromised by this side effect of the transgene. It is expected that the vectors containing long genomic DNA regions around the WAP gene will be able to reduce position effects leading to leaky expression of the transgenes.

Data of Table 8.1 show the list of the genes which have directed the secretion of pharmaceutical proteins in rabbit milk. This list is not complete as a certain number of the projects in course are kept confidential.

8.5 Quality of the Proteins Secreted in Milk

A large proportion of the proteins to be used as medicaments have a complex structure. They must be folded as the native corresponding proteins. This implies the formation of appropriate disulphide bridges. The mature proteins may be formed by the association of several subunits. They may be N- or O-glycosylated, γ-carboxylated, sulphated or phosphorylated. They may also be cleaved at specific sites to eliminate signal peptides or propeptides but also to generate subunits.

The carbohydrate chain must contain a terminal sialic acid, preferably under the NANA (N-acetylneuraminic acid) form to be stable in blood. In a few cases only, the carbohydrate content of recombinant proteins secreted in milk has been described in detail. This was the case for human α-glucosidase (Bijvoet et al. 1999) and inhibitor C1 (Koles et al. 2004a,b) which was produced at a very high concentration (up to 12 mg/ml). This protein was not completely glycosylated, possibly due to the saturation of the cellular machinery involved in glycosylation. This reduced very significantly the half life of the protein injected into patients. Interestingly, this protein contained only traces of the NGNA (N-glycosylneuraminic acid) which

Table 8.1 List of the major recombinant pharmaceutical proteins prepared in transgenic rabbit milk

Recombinant expressed protein	Promoter	Foreign gene	Expression level	Specific features	Reference
Interleukin 2 (IL2)	Rabbit β-casein	Genomic DNA	0.0005 g/l		Buhler et al., 1990
Human growth hormone	Mouse WAP	Genomic DNA	0.05 g/l	hGH in the serum	Limonta et al., 1995
Human insulin-like growth factor	Bovine αs1-casein expression cassette	cDNA	1 g/l	Upregulation of IGFB-2, increased milk yield	Brem et al., 1994 Wolf et al., 1997, Zinovieva et al., 1998
Bovine chymosin	Bovine αs1-casein expression cassette	Prochymosin gene	10 g/l	Healthy animals and no unwanted clotting	Brem et al., 1995
Human erythropoietin	Rabbit WAP	cDNA	0.0003 g/l		Rodriguez et al., 1995
Human erythropoietin	Bovine β-lactoglobulin	cDNA (fusion protein of erythropoietin with β-lactoglobulin to avoid biological activity in the secreting animal	0.05 g/l	Correct glycosylation, transient increase in hematocrit values of lactating females	Korhonen et al., 1997
Human EC superoxide dismutase	Mouse WAP	cDNA EC SOD	3 g/l	Not fully glycosylated, biologically active	Strömqvist et al., 1996

Salmon calcitonin	Ovine β-lactoglobulin	cDNA (fusion protein of salmon calcitonin with human-lactoglobulin)	2.1 g/l	First evidence of C-terminal amidation by the mammary gland	McKee et al., 1998
Human Nerve growth factor beta	Bovine αs1-casein expression cassette	cDNA	0.25 g/ml		Coulibaly et al., 1999
Equine chorionic gonadotropin	Rabbit WAP	β/α chain	0.022 g/l	Rapid plasma clearance no *in vivo* biological activity	Galet et al., 2001
Bovine follicle stimulating hormone	Bovine αs1-casein expression casette	cDNA α and β chains	0.005 g/l	Two double transgenic rabbit lines, full *in vitro* biological activity	Coulibaly et al., 2002
Human C1 inhibitor			12 g/l	Replacement therapy in hereditary angioderma	Koles et al., 2004a,b
Human factor VIII	Mouse WAP	cDNA	Up to 60% of the level in human blood		Chrenek et al., 2007

is never present in native human protein as the gene involved in the formation of NGNA is mutated and no more functional in higher primates. This is a clear advantage over proteins produced in goat milk such as human anti-thrombin III which contained almost as many NGNA as NANA (Edmunds et al. 1998). This might modify the bioactivity of the protein and potentially induce the formation of anti-NGNA antibodies in patients. Interestingly also, the human inhibitor C1 was poorly fucosylated. This may be important for antibodies which have a much higher ADCC reaction when they are not fucosylated (Jeffris 2006). The recombinant antibodies secreted in milk may thus have a better anti-tumour activity than those secreted by CHO cells and which are fucosylated.

The same conclusions were drawn with a human coagulation blood factor which was however better glycosylated. Moreover, this factor was cleaved and γ-carboxylated as the native protein (Chtourou et al. patent WO2008015339).

8.6 The Specific Properties of Rabbits as Living Fermentor

Rabbit is a relatively small animal and it cannot produce as much protein as a goat or a cow. Yet, between 10 and 15l of milk can be obtained from a female per year using a milking machine. This means that a herd of a few hundreds rabbits can produce several kilograms of a recombinant protein per year.

Transgenic rabbits can be obtained with no particular difficulty and this species is very prolific. No more than seven months are required between the beginning of DNA microinjection and the first collection of milk from a founder transgenic female. Rabbits can be bred easily and at a relatively low cost in specific pathogen-free conditions. Moreover, rabbits are not sensitive to prion diseases and they do not transfer pathogens dangerous for humans.

Breeding a SPF production herd is very well known and easy task as it is done for husbandry containing thousands of rabbit every day to produce polyclonal antibodies (Thymoglobulin®). A milk production herd of 500 rabbits with a middle range production rate (4 g/l) will be enough to produce around 10 kg of final product.

Rabbit milking may seem laborious but it is facilitated by the use of well-adapted milking machine and trained husbandry technicians.

8.7 Conclusion and Perspectives

Various types of proteins have been prepared in rabbit milk successfully. These proteins are either enzymes, vaccines, hormones, growth factors or blood factors. About 550 monoclonal antibodies are under study and a proportion of them will have to be prepared for patients. In addition, several hundreds of other proteins could be validated for being used as medicaments. Some experts postulated that in the coming years, all the systems for the preparation of recombinant proteins might be insufficient to meet the demand of patients (Houdebine 2002). Rabbits appear the most appropriate for preparing a few kilograms of a protein per year. This represents the needed amount for most pharmaceutical proteins. Rabbits offer a relatively low investment cost even if milk collection may be more costly than in cows or goats. In case the production of a protein from rabbit milk becomes insufficient to meet the demand of the market, larger animals harbouring the same gene construct can be generated.

Although the tools to generate transgenic rabbits and to obtain a reliable high level of recombinant proteins in milk have become efficient, they may be and they will be improved in the coming years. This will make rabbits still more attractive to prepare pharmaceutical proteins.

References

Bijvoet AG, Van Hirtum H, Kroos MA, Van de Kamp EH, Schoneveld O, Visser P, Brakenhoff JP, Weggeman M, van Corven EJ, Van der Ploeg AT, Reuser AJ. (1999) Human acid alpha-glucosidase from rabbit milk has therapeutic effect in mice with glycogen storage disease type II. Hum Mol Genet 8: 2145–2153.

Bösze Zs, Houdebine LM. (2006) Application of rabbits in biomedical research: review. World Rabbit Sci 14: 1–14.

Brem G, Hartl P, Besenfelder U, Wolf E, Zinovieva N, Pfaller R. (1994) Expression of synthetic cDNA sequences encoding human insulin-like growth factor-1 (IGF-1) in the mammary gland of transgenic rabbits. Gene 149: 351–355.

Brem G, Besenfelder U, Zinovieva N, Seregi J, Solti L, Hartl P. (1995) Mammary gland specific expression of chymosin constructs in transgenic rabbits. Theriogenology 43: 175–175.

Buelow R, van Schooten W. (2006) The future of antibody therapy. Ernst Schering Found Symp Proc 4: 83–106.

Buhler TA, Bruyere T, Went DF, Stranzinger G, Burki K. (1990) Rabbit beta-casein promoter directs secretion of human interleukin-2 into the milk of transgenic rabbits. Biotechnology (NY) 8: 140–143.

Chesné P, Adenot PG, Viglietta C, Baratte M, Boulanger L, Renard JP. (2002) Cloned rabbits produced by nuclear transfer from adult somatic cells. Nat Biotechnol 20: 366–369.

Chrenek P, Ryban L, Vetr H, Makarevich AV, Uhrin P, Paleyanda RK, Binder BR. (2007) Expression of recombinant human factor VIII in milk of several generations of transgenic rabbits. Transgenic Res 16: 353–361.

Coulibaly S, Besenfelder U, Fleischmann M, Zinovieva N, Grossmann A, Wozny M, Bartke I, Tögel M, Müller M, Brem G. (1999) Human nerve growth factor beta (hNGF-beta): mammary gland specific expression and production in transgenic rabbits. FEBS Lett 444: 111–116.

Coulibaly S, Besenfelder U, Miller I, Zinovieva N, Lassnig C, Kotler T, Jameson JL, Gemeiner M, Müller M, Brem G. (2002) Expression and characterization of functional recombinant bovine follicle-stimulating hormone (boFSHalpha/beta) produced in the milk of transgenic rabbits. Mol Reprod Dev 63: 300–308.

Edmunds T, Van Patten SM, Pollock J, Hanson E, Bernasconi R, Higgins E, Manavalan P, Ziomek C, Meade H, McPherson JM, Cole ES. (1998) Transgenically produced human antithrombin: structural and functional comparison to human plasma-derived antithrombin. Blood 91: 4561–4571.

Galet C, Le Bourhis CM, Chopineau M, Le Griec G, Perrin A, Magallon T, Attal J, Viglietta C, Houdebine LM, Guillou F. (2001) Expression of a single betaalpha chain protein of equine LH/CG in milk of transgenic rabbits and its biological activity. Mol Cell Endocrinol 174: 31–40.

Giraldo P, Rival-Gervier S, Houdebine LM, Montoliu L. (2003) The potential benefits of insulators on heterogonous constructs in transgenic. Transgenic Res 12: 751–755.

Gordon K, Lee E, Vitale JA, Smith AE, Westphal H, Hennighausen L. (1987) Production of human tissue plasminogen activator in transgenic mouse milk. Bio/Technol 5: 1183–1187.

Hammer RE, Pursel VG, Rexroad CE Jr, Wall RJ, Bolt DJ, Ebert KM, Palmiter RD, Brinster RL. (1985) Production of transgenic rabbits, sheep and pigs by microinjection. Nature 315: 680–683.

Han JY. (2008) Germ cells and transgenesis in chicken. Comp Immunol Microbiol Infect Dis 32:61–80.

Houdebine LM. (2002) Antibody manufacture in transgenic animals, comparisons with other systems. Curr Opin Biotechnol 13: 625–629.

Houdebine LM. (2007) Transgenic animal models and target validation. Meth Mol Biol 360: 163–202.

Houdebine LM. (2008) Methods to generate transgenic animals in ethics of science and technology assessment. Springer. Heildelberg/Berlin/New York 34: 31–48.

Houdebine LM. (2009) Production of pharmaceutical proteins by transgenic animals. Comparative Immunology, Microbiology & Infectious Diseases 32: 107–121.

Jeffris R. (2006) A sugar switch for anti-inflammatory antibodies Nat Biotechnol 24: 1230–1231.

Korhonen VP, Tolvanen M, Hyttinen JM, Uusi-Oukari M, Sinervirta R, Alhonen L, M, Jänne OA, Jänne J. (1997) Expression of bovine beta-lactoglobulin/human erythropoietin fusion protein in the milk of transgenic mice and rabbits. Eur J Biochem. 245: 482–489.

Koles K, van Berkel PH, Pieper FR, Nuijens JH, Mannesse ML, Vliegenthart JF, Kamerling JP. (2004a) N- and O-glycans of recombinant human C1 inhibitor expressed in the milk of transgenic rabbits. Glycobiology 14: 51–64.

Koles K, van Berkel PH, Mannesse ML, Zoetemelk R, Vliegenthart JF, Kamerling JP. (2004b) Influence of lactation parameters on the N-glycosylation of recombinant human C1 inhibitor isolated from the milk of transgenic rabbits. Glycobiology 14: 979–986.

Lillico SG, Sherman A, McGrew MJ, Robertson CD, Smith J, Haslam C, Barnard P, Radcliffe PA, Mitrophanous KA, Elliot EA, Sang HM. (2007) Oviduct-specific expression of two therapeutic proteins in transgenic hens. Proc Natl Acad Sci U S A 104: 1771–1776.

Limonta J, Pedraza A, Rodriguez A, Freyre FM, Barral AM, Castro FO, et al. (1995) Production of active anti-CD6 mouse/human chimeric antibodies in the milk of transgenic mice. Immunotech 1: 107–113.

Massoud M, Bischoff R, Dalemans W, Pointu H, Attal J, Schultz H, Clesse D, Stinnakre MG, Pavirani A, Houdebine LM. (1991) Expression of active recombinant human alpha 1-antitrypsin in transgenic rabbits. J Biotechnol 18: 193–203.

Massoud M, Attal J, Thépot D, Pointu H, Stinnakre MG, Théron MC, Lopez C, Houdebine LM. (1996) The deleterious effects of human erythropoietin gene driven by the rabbit whey acidic protein gene promoter in transgenic rabbits. Reprod Nutr Dev 36: 555–563.

McKee C, Gibson A, Dalrymple M, Emslie L, Garner I, Cottingham I. (1998) Production of biologically active salmon calcitonin in the milk of transgenic rabbits. Nat Biotechnol 16: 647–651.

Millot B, Fontaine ML, Thépot D and Devinoy E. (2001) A distal region, hypersensitive to DNase I, plays a key role in regulating rabbit whey acidic protein gene expression. Biochem J 359: 557–565.

Moreira PN, Pozueta J, Pérez-Crespo M, Valdivieso F, Gutiérrez-Adán A, Montoliu L. (2007) Improving the generation of genomic-type transgenic mice by ICSI. Transgenic Res 16: 163–168.

Palmiter RD, Brinster RL, Hammer RE, Trumbauer ME, Rosenfeld MG, Birnberg NC, Evans RM. (1982) Dramatic growth of mice that develop from eggs microinjected with metallothionein-growth hormone fusion genes. Nature 300: 611–615.

Pera MF, Hasegawa K. (2008) Simpler and safer cell reprogramming. Nat Biotechnol 26: 59–60.

Rival S, Attal J, Delville-Giraud C, Yerle M, Laffont P, Rogel-Gaillard C, Houdebine LM. (2001) Cloning, transcription and chromosomal localization of the porcine whey acidic protein gene and its expression in HC11 cell line. Gene 267: 37–47.

Rival-Gervier S, Viglietta C, Maeder C, Attal J, Houdebine LM. (2002) Position-independent and tissue-specific expression of porcine whey acidic protein gene from a bacterial artificial chromosome in transgenic mice. Mol Reprod Dev 63: 161–167.

Rival-Gervier S, Maeder C,Viglietta C, Prince S and Houdebine LM. (2003) Effect of 5'HS4 insulator on rabbit WAP gene action in transgenic mice. Transgenic Res 12: 12723–12730.

Rodriguez A, Castro FO, Aguilar A, Ramos B, Del Barco DG, Lleonart R, De la Fuente J. (1995) Expression of active human erythropoietin in the mammary gland of lactating transgenic mice and rabbits. Biol Res 28: 141–153.

Saidi S, Rival-Gervier S, Daniel-Carlier N, Thépot D, Morgenthaler C, Viglietta C, Prince S, Passet B, Houdebine LM and Jolivet G. (2007) Distal control of the pig whey acidic protein (WAP) locus in transgenic mice. Gene 401: 97–107.

Shen W, Li L, Pan Q, Min L, Dong H, Deng J. (2006) Efficient and simple production of transgenic mice and rabbits using the new DMSO-sperm mediated exogenous DNA transfer method. Mol Reprod Dev 73: 589–594.

Shinohara ET, Kaminski JM, Segal DJ, Pelczar P, Kolhe R, Ryan T, Coates CJ, Fraser MJ, Handler AM, Yanagimachi R, Moisyadi S. (2007) Active integration: new strategies for transgenesis. Transgenic Res 16: 333–339.

Smith K, Spadafora C. (2005) Sperm mediated gene transfer: applications and implications. BioEssays 27: 551–562.

Soler E, Le Saux A, Guinut F, Passet B, Cohen R, Merle C, Charpilienne A, Fourgeux C, Sorel V, Piriou A, Schwartz-Cornil I, Cohen J, and Houdebine LM. (2005) Production of two vaccinating recombinant rotavirus proteins in the milk of transgenic rabbits. Transgenic Res 14: 833–844.

Strömqvist M, Houdebine LM, Andersson JO, Edlund A, Johansson T, Viglietta C, Puissant C, Hansson L. (1996) Recombinant human extracellular superoxide dismutase produced in milk of transgenic rabbits. Transgenic Res 6: 271–278.

Van de Lavoir MC, Diamond JH, Leighton PA, Mather-Love C, Heyer BS, Bradshaw R, Kerchner A, Hooi LT, Gessara TM, Swanberg SE, Delany ME, Etches RJ. (2006) Germline transmission of genetically modified primordial germ cells. Nature 441: 766–769.

Weidle UH, Lenz H, Brem G. (1991) Genes encoding a mouse monoclonal antibody are expressed in transgenic mice, rabbits and pigs. Gene 98: 185–191.

Wolf E, Jehle PM, Weber MM, Sauerwein H, Daxenberger A, Breier BH, Besenfelder U, Frenyo L, Brem G. (1997) Human insulin-like growth factor I (IGF-I) produced in the mammary glands of transgenic rabbits: yield, receptor binding, mitogenic activity, and effects on IGF-binding proteins.Endocrinology 138: 307–313.

Yong HY, Hao Y, Lai L, Li R, Murphy CN, Rieke A, Wax D, Samuel M, Prather RS. (2006) Production of a transgenic piglet by a sperm injection technique in which no chemical or physical treatments were used for oocytes or sperm. Mol Reprod Dev 73: 595–599.

Zinovieva N, Lassnig C, Schams D, Besenfelder U, Wolf E, Müller S, Frenyo L, Seregi J, Müller M, Brem G. (1998) Stable production of human insulin-like growth factor 1 (IGF-1) in the milk of hemi- and homozygous transgenic rabbits over several generations. Transgenic Res 7: 437–447.

Chapter 9
Derivation and Characterization of Rabbit Embryonic Stem Cells: A Review

Elen Gócza and Zsuzsanna Bősze

Abstract In the first part of this chapter the different types of pluripotent stem cells are described in general: embryonal carcinoma cells, mouse and human embryonic stem cells, germ cells, epiblast stem cells and induced pluripotent cells. The methods used for isolation of rabbit embryonic stem like cells and rabbit primordial germ cells are detailed in the second part, including the species specific factors playing role in the maintenance of pluripotency. Detection of pluripotency markers both at mRNA and protein levels is an important tool in embryonic stem cell characterization. In vitro differentiation, teratoma formation and chimera forming ability are all inevitable tools to characterize embryonic stem cells. Finally the future potential of a truly validated widely available rabbit embryonic stem cell line is highlighted.

Keywords Embryonic stem cells, pluripotency, chimera, in vitro differentiation

9.1 Introduction

Embryonic stem cells (ESC) are derived from the inner cell mass (ICM) of blastocyst stage embryos, can be maintained indefinitely in vitro, while retaining the ability to form all cell types of the body following normal embryogenesis. Embryonic stem cells serve as source material for the production of genetically modified animals and they can potentially used in tissue and cell therapy. Although ESC lines derived from mice, monkeys and humans, and are used extensively, the development of such lines from other mammals, like rabbit would be still valuable. The derivation of human embryonic stem cells (hESC) 10 years ago was based on decades of research on mouse embryonic carcinoma cells (ECC) and embryonic stem cells. Similarly, the recent derivation of mouse and human induced pluripotent stem cells (iPS) depended on prior studies on mouse and human embryonic stem cells (Yu and Thomson 2008).

Z. Bősze (✉)
Agricultural Biotechnology Center, H-2100 Gödöllő, Szent-Györgyi A. St. 4, Hungary
e-mail: bosze@abc.hu

L.-M. Houdebine and J. Fan (eds.), *Rabbit Biotechnology: Rabbit Genomics, Transgenesis, Cloning and Models*,

Due to the efforts which were made earlier (Cole et al. 1966; Graves and Moreadith 1993; Schoonjans et al. 1996) and recently (Fang et al. 2006; Wang et al. 2007) to derive rabbit ESC lines a bulk of information was already collected. Here we decribe the history of rabbit embryonic stem cells line (rabESC) establishment, compare the methods used in the light of recent developments in other species and highlight the future potential hiden in an improved, widely available rabbit ESC line.

9.2 Pluripotent Stem Cells of Different Origins

9.2.1 Embryonic Carcinoma Cells (ECC)

Historically the pluripotent stem cell research began with establishment of teratocarcinomas in the 1950s. Teratocarcinomas are malignant germ cell tumors that comprise an undifferentiated EC component and a differentiated component that can include all three germ layers. Kleinsmith and Pierce (1964) demonstrated that a single EC cell is capable of both unlimited self-renewal and multilineage differentiation, which was defined as pluripotent stem cell. Mouse EC cell lines that could be stably propagated in vitro were established in the early 1970s (Kahan and Ephrussi 1970). Indeed, some EC cell lines are able to contribute to various somatic cell types upon injection into mouse blastocysts (Brinster 1974). Human EC cells were subsequently derived (Hogan et al. 1977). Human EC cells are highly aneuploid, which likely accounts for their inability to differentiate into a wide range of somatic cell types, and limits their utility as an in vitro model of human development.

9.2.2 Mouse Embryonic Stem Cells (mESC)

The first mouse ES cell lines were derived from the ICM of mouse blastocysts using culture conditions previously used for mouse EC cells (Evans and Kaufman 1981; Martin 1981). ES cells were differentiated into a wide variety of cell types in vitro and formed teratocarcinomas when injected into mice. ES cells can contribute at a high frequency to a variety of tissues in chimeras, including germ cells, thus providing a practical way to introduce modifications to the mouse germline (Bradley et al. 1984). Notably, the scientists, who derived these cells for the first time and used them to identify the function of genes have been celebrated with the Nobel price in 2007 (Doetschman et al. 1987; Evans and Kaufman 1981; Thomas and Capecchi 1987).

9.2.3 Human Embryonic Stem Cells (hESC)

In the mid-1990s, ES cell lines were derived from two nonhuman primates: the rhesus monkey and the common marmoset (Thomson et al. 1995, 1996). Experience with those primate ES cell lines permitted the subsequent derivation of human ES cell lines (Thomson et al. 1998). Human ES cells are karyotypically normal and, even after prolonged undifferentiated proliferation, maintain the developmental potential to contribute to advanced derivatives of all three germ layers. Mitotically inactivated mouse fibroblast feeder layers and serum-containing medium were used in the initial derivation of human ES cells. Mouse embryonic fibroblast feeder layer support both mouse and human ES cells, but LIF and its related cytokines fail to support human or nonhuman primate ES cells in serum-containing media that supports mouse ES cells (Daheron et al. 2004; Humphrey et al. 2004; Thomson et al. 1998). Consistent with this observation, human ES cells do not express or only at very low level the critical components of the LIF pathway (Brandenberger et al. 2004). In contrast to mouse ES cells, FGF and TGF/Activin/Nodal signaling are of central importance to the self-renewal of human ES cells, making human ES cells similar to the recently described mouse epiblast-derived stem cells (Brons et al. 2007; Tesar et al. 2007). Basic FGF (bFGF) allows the clonal growth of human ES cells on fibroblasts in the presence of serum replacement (Amit et al. 2000). At higher concentrations, bFGF allows feeder independent growth of human ES cells cultured in the same serum replacement (Levenstein et al. 2006; Li et al. 2005; Xu et al. 2005).

9.2.4 Pluripotent Cell Lines Derived from Germ Cells (EG cells)

It was 15 years ago, when pluripotent stem cells (embryonic germ cells or EG cells) were successfully derived from primordial germ cells (PGCs) (Matsui et al. 1992; Resnick et al. 1992). Derivation of mouse EG cells requires a combination of stem cell factor (SCF), LIF and FGF in the presence of feeder layer. EG cells are morphologically indistinguishable from mouse ES cells, express typical ES cell markers and upon blastocyst injection, contribute extensively to chimeric mice including germ cells (Labosky et al. 1994; Stewart 1994). Recently multipotent germline stem cells (mGSCs) have been derived from both neonatal and adult mouse testis (Guan et al. 2006; Kanatsu-Shinohara et al. 2004). mGSCs express typical mouse ES cell-specific markers, differentiate into multiple lineages in vitro, form teratomas, and contribute to chimeras upon injection into blastocysts. Nevertheless, mGSCs have an epigenetic status distinct from both ES cells and germline stem cells (Kanatsu-Shinohara et al. 2004).

Derivation of human EG cells was first reported in 1998 (Shamblott et al. 1998), but their long-term proliferative potential appears to be limited (Turnpenny et al. 2003). Early passage human EG cells have been reported to differentiate into multiple lineages in vitro, but to date no human EG cell lines have been reported to form teratomas (Shamblott et al. 1998).

In avian embryos, PGCs are first identified in an extra-embryonic region, the germinal crescent, after approximately 18 h of incubation. After 50–55 h of development, PGCs migrate to the gonad and subsequently produce functional sperm and oocytes. Chicken PGCs can be isolated, cultured and genetically modified, while maintaining their commitment to the germ line. Furthermore chicken PGCs can be induced in vitro to differentiate into embryonic germ cells that contribute to somatic tissues (van de Lavoir et al. 2006).

9.2.5 Epiblast Stem Cells (EpiSCs)

A novel type of pluripotent stem cell lines, called epiblast stem cells (EpiSCs) have been established from epiblasts isolated from E5.5 to E6.5 post-implantation mouse embryos. EpiSCs differ significantly from mouse ES cells but share some features with human ES cells (Brons et al. 2007; Tesar et al. 2007). EpiSCs derivation failed in the presence of LIF and/or BMP4, but similarly to human ES cells FGF and Activin/Nodal signalling appeared critical for EpiSC derivation. Comparison of those epiblast cell lines to human ES cells suggest, that human ES cells are more equivalent to the early postimplantation epiblasts, than to the ICM progenitors (Rossant 2008).

9.2.6 Induced Pluripotent Stem Cells

It was a real breakthrough, when the Yamanaka group (Takahashi and Yamanaka 2006) identified Oct4, Sox2, c-Myc, and Klf4 as sufficient to reprogram mouse fibroblasts to cells closely resembling mouse ES cells. These results were rapidly confirmed and extended in mouse material (Maherali et al. 2007; Okita et al. 2007; Wernig et al. 2007) and eventually successfully applied to human material (Lowry et al. 2008; Park et al. 2008; Takahashi and Yamanaka 2006; Yu et al. 2006). It is generally accepted by now, that Oct4, Sox2 and Nanog are essential factors, which influence the efficiency of reprogramming (Yu et al. 2007).

9.2.7 Embryo Derived Stem Cell Lines in Domestic Species

Many reports described cell lines derived in domestic species, which presented several important features typical of embryonic stem cells (ESCs). Putative ESCs in sheep, pig and cow were published in 1990–1991 (Notarianni et al. 1991; Piedrahita et al. 1990a, b).

Those first generation pig and rabbit ESC lines were able to form chimeras (Schoonjans et al. 1996; Wheeler 1994), however the integration of embryo derived cells into the germline of any domestic species was not demonstrated (Brevini et al. 2008; Keefer et al. 2007).

9.3 Rabbit Embryonic Stem Cells

Laboratory rabbits have long been used in biomedical research. Transgenic rabbit models created through "traditional" microinjection method widely used as experimental models of human diseases such as atherosclerosis, myocardial infarction, hypercholesterolemia, hypertension, bone and cartilage disorders (Bosze and Houdebine 2006; Shiomi and Fan 2008). Rabbit ES cells would be invaluable both for creating second generation transgenic models of human diseases using gene-targeted technology and for testing stem cell therapies for human applications. Although many attempts have been made to derive ES cell lines from rabbits, to date germ line competent rabbit ESC lines are not available (Cole et al. 1966; Fang et al. 2006; Graves and Moreadith 1993; Honda et al. 2008; Schoonjans et al. 1996; Wang et al. 2007).

9.3.1 Methods Used for Derivation of Rabbit ES Like Cell Lines

Several reports on rabbit ES like cell lines derivation have been published since the first attempt in 1966 (Cole et al.). The pioneering work by Graves and Moreadith (1993) described the derivation and initial characterisation of putative embryonic stem cell lines from preimplantation rabbit embryos. Two principal cell types emerged following serial passage of explanted embryos, and each had subsequently gave rise to immortalised cell lines. One cell type had identical morphology to primary outgrowths of trophectoderm, it was strictly feeder-cell dependent and spontaneously formed trophectodermal vesicles at high cell density. Pluripotent ES cells, the second cell type, derived from the inner cell mass maintained of a predominantly normal karyotype through serial passage and could form embryoid bodies. These ES cell lines were further characterised and their capacity to contribute to formation of adult, fertile animals upon injection into recipient blastocysts was demonstrated. The efficiency of chimera formation was low (5% of live born), but the degree of chimerism, as assessed by coat colour contribution was 10–50%, relatively high (Schoonjans et al. 1996).

The special feature of rabbit blastocysts, that a strong zona pellucida and a very thick outer mucin coat surround them. In vivo, the hatching blastocysts escape both from zona pellucida and the mucin coat through a hole made by trophoblastic knobs (Steer 1970). Therefore, in the in vitro experiments, before placing blastocysts into the culture medium, the zona pellucida should be removed. It can be reached either by treating the embryos with 0.5% pronase (Sigma, USA) in PBS (Catunda et al. 2008), or with 0.5% protease (Sigma-Aldrich) followed by incubation of the zona-free blastocysts with 0.05% trypsin (Sigma-Aldrich) 0.008% EDTA (Sigma-Aldrich) for 3–5 min. Thereafter, the inner cell mass (ICM) could be mechanically separated from the trophectoderm (Wang et al. 2007). The main steps of ES cell line establisment are described in Fig. 9.1. However, those exogenous, enzymatic treatments may damage blastocysts. Honda and his colleagues (2008) mimicked natural blastocyst hatching: a hole was made in the zona near the space between two blastomeres using a Piezo micromanipulator (Fig. 9.2). Blastocysts escaped from the zona pellucida through the hole, when medium was introduced into the perivitelline space (Yamagata et al. 2002). With this method, the zona pellucida-free blastocysts were collected less harmfully and used immediately for ES cell derivation.

Feeder cell layers are extensively used for derivation and maintenance of ES cells or ES-like cells in many species. Feeder cells, mostly of murine origin are

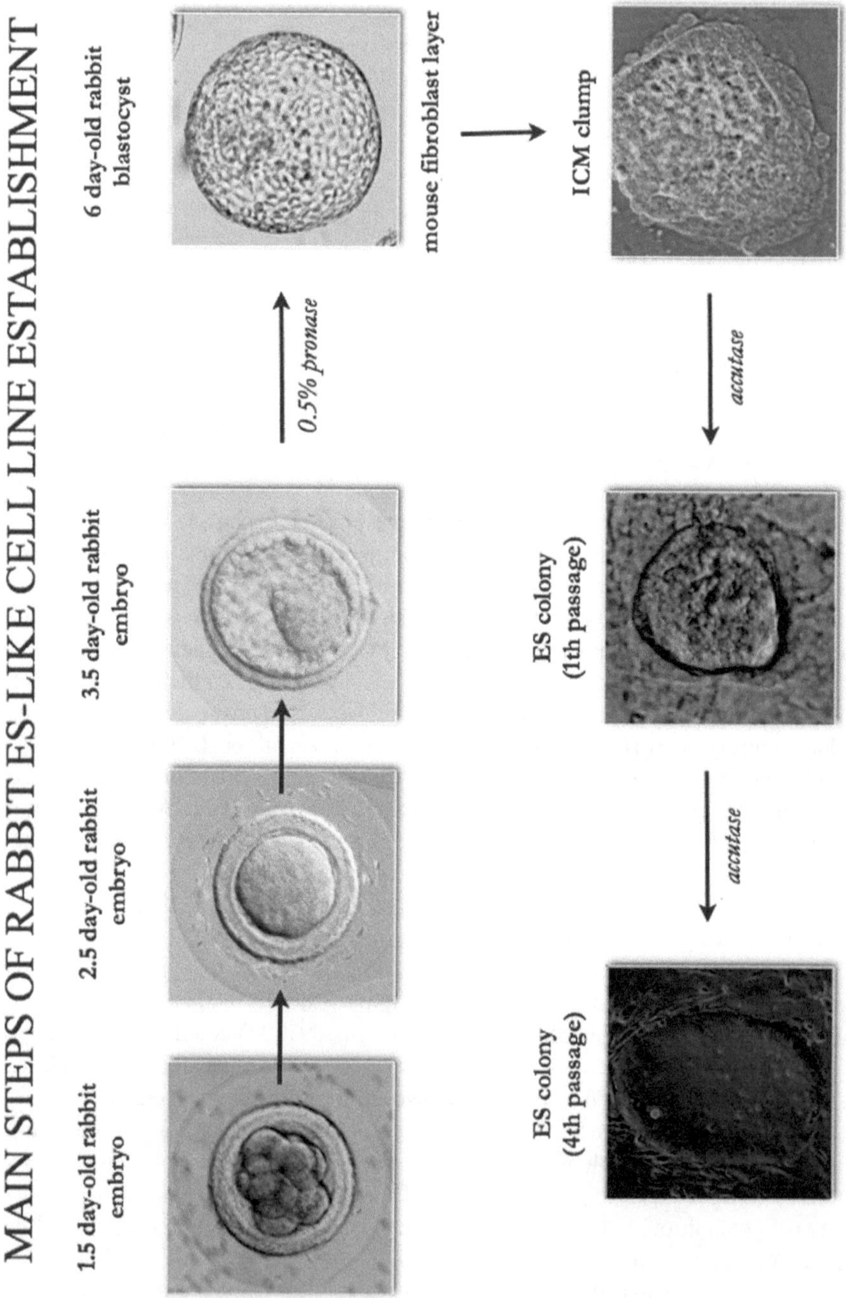
MAIN STEPS OF RABBIT ES-LIKE CELL LINE ESTABLISHMENT
1.5 day-old rabbit embryo
2.5 day-old rabbit embryo
3.5 day-old rabbit embryo
0.5% pronase
6 day-old rabbit blastocyst
mouse fibroblast layer
ICM clump
accutase
ES colony (1th passage)
accutase
ES colony (4th passage)

thought to support ES cells by supplying nutrients and growth factors in a paracrine fashion and by providing scaffolds for ES cell colonisation (Choo et al. 2006; Stacey et al. 2006). As feeder layer mouse embryonic fibroblast (MEF) cell layer derived from 13.5-day-old foetuses of CD1 (Catunda et al. 2008; Honda et al. 2008) or 129SV origin (Wang et al. 2007) are used. Honda et al. (2008) found that high density MEFs repressed the proliferation of ES cells and the optimal efficiency for ES cell establishment was observed at 6×10^3cells/cm^2. Nevertheless, feeder cells are apparently indispensable for the maintenance of the pluripotency of rabbit ES cells as they transform into fibroblast like cells in the absence of feeder cells.

The rabbit ESC cultivation medium was developed with slight modification of the medium adapted by Vassilieva et al. (2000) for long-term cultivation of rat embryonic stem-like cell lines. She used a modified mouse ES cell culture medium which contained DMEM: IMDM (1:1, 4.5 g/l glucose; GIBCO) supplemented with 15% of fetal calf serum (ES cells qualified; GIBCO) and basic additives containing 1% nonessential amino acids (stock solution 1:100, NEAA, GIBCO), 2 mM L-glutamine (stock solution 1:100, GIBCO), 1% nucleosides (stock solution 1:100 Sigma), 15 mg/ml human transferrin (Boehringer Mannheim), 30 nM sodium selenite (Sigma), 450 mM α-monothioglycerol (MTG, Sigma), penicillin/streptomycin (50 IU/50 mg, GIBCO), 20 ng/ml human LIF and 20 ng/ml basic fibroblast growth factor (bFGF; Strathmann Biotech, Germany). Catunda et al. (2008) used rat LIF in cell culture medium instead of human LIF. Fang and his collages (2006) used 78% DMEM/F12, 20% knockout serum replacement (Invitrogen, Carlsbad, CA) with 4–8 ng/ml human recombinant bFGF (basic fibroblast growth factor). Wang and his group (2007) used a Dulbecco's modified Eagle's medium (high glucose, without sodium pyruvate; Invitrogen Corporation) with standard supplements and 15% defined fetal bovine serum (HyClone, Logan, UT) but without LIF. The culture medium used by Honda et al. (2008) consisted of 78% DMEM/F-12 (Invitrogen, Carlsbad, CA, USA), 20% knockout serum replacement (KSR; Invitrogen), 2 mM GlutaMax (Invitrogen), 1% nonessential amino acids, 0.1 mM β-mercaptoethanol, 10^3 U/ml ESGRO LIF and 8 ng/ml human bFGF (Wako, Osaka, Japan). Reduced expression of Oct4 in rabbit ES cells were observed, when rabbit ES cells were cultured in the presence of serum (FBS), and reduced and uneven distribution of alkaline-phosphatase (AP) positive cells were noted by Honda et al. (Fig. 9.3).

Fig. 9.1 Rabbit eight-cell stage embryos were collected at 1.5 days postcoitum and were cultured in vitro for 2 days. From the 3.5-day-old blastocysts, the zona pellucida was removed by treating with 0.5% pronase (Sigma) in PBS and after washing the blastocysts in culture medium they were placed onto mitomycin C (10 mg/ml, Sigma) inactivated mouse primary embryonic fibroblasts in 0.1% gelatin (Sigma) coated wells of tissue-culture plates (Nunc). Hatched embryos were allowed to attach onto the feeder layer and formed embryonic outgrowths. Undifferentiated cells of embryonic outgrowths were placed into accutase (Sigma) drops and were mechanically disaggregated. Aggregates of 20–50 cells were seeded onto fresh feeder cells. Colonies presenting a typical morphology of compacted ES cells were selected and subcultured every week using accutase to disaggregate cells (Adapted from Catunda et al. (2008) with the permission from Mary Ann Liebert Inc.)

RABBIT ES LIKE CELL LINE ESTABLISHMENT AFTER ARTIFICIAL ZONA SHEDDING USING A PIEZO MICROMANIPULATOR

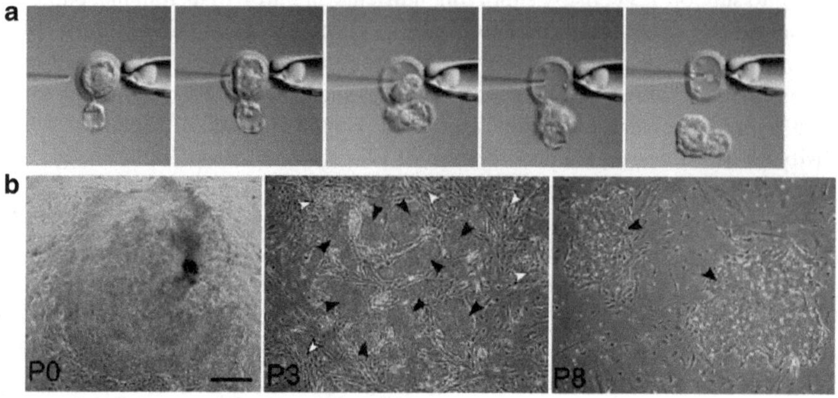

Fig. 9.2 Derivation of primary outgrowth from rabbit blastocyst. (**a**) Artificial zona shedding of a blastocyst using a Piezo micromanipulator. The blastocyst was smoothly flushed through a slit by introducing medium into the perivitelline space from the opposite side. (**b**) The appearance of cell colonies at different stages, from the outgrowth of ICM cells to the emergence of stable ES colonies. P0: A primary rabbit ES cell colony grown on feeder cells 7 days after plating. P3: Mixed populations of stem-cell-like (black arrowheads) and differentiating (white arrowheads) colonies at passage 3. P8: Only stem-cell-like colonies remained after the eighth passage (black arrow). Scale bar: 50 μm (Adapted from Honda et al. (2008) with the permission of Ingenta Connect)

EFFECT OF CULTURE MEDIUM ON RABBIT ES-LIKE CELL LINE ESTABLISHMENT

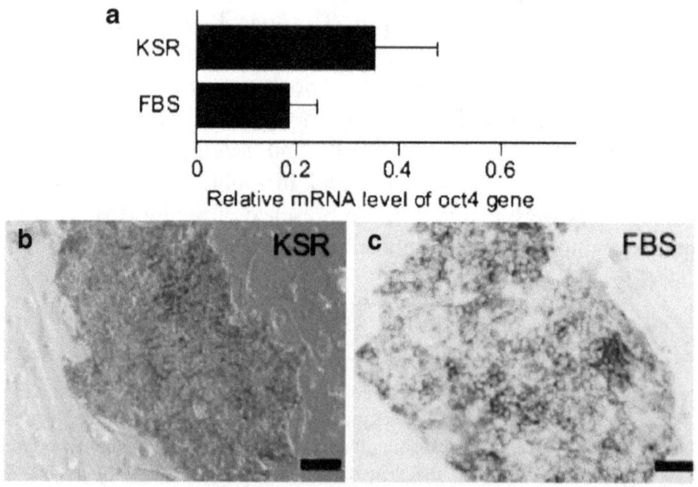

Fig. 9.3 (**a**) Relative expression levels of rabbit Oct4 in the presence of KSR or FBS. Reduced expression was observed when rabbit ES cells were cultured in the presence of serum. (**b–c**) Alkaline phosphatase activity of rabbit ES cell colonies cultured in the presence of KSR or FBS. The reduced and uneven distribution of alkaline-phosphatase-positive cells is noted in the FBS group (Adapted from Honda et al.(2008) with the permission of Ingenta Connect)

Rabbit ICMs and ESC-like cells are very sensitive to enzyme digestion and mechanical damages, because they would rapidly differentiate or die if exposed to higher concentrations of enzyme, longer time, or mechanical damage, especially during the first five passages. Therefore great varieties were published in this step of establishment. The ESC-like cell colonies were mechanically divided into small clumps by micropipetting, followed by treatment of 5 mg/ml dispase for several minutes and further propagated in clumps of five to approximately 50 cells on fresh MEFs in Wang's publication (2007). Catunda and her co-workers (2008) used accutase (Sigma) to pass the rabbit ES like cell colonies. 0.05% trypsin for 1 min at room temperature and mechanical disaggregating of small clumps into single cells was used by Honda et al. (2008). In most cases rabbit ESC lines were frozen in liquid nitrogen using freezing solution consisting of 10% dimethyl sulfoxide (Sigma) and 90% FCS.

9.3.2 Isolation and Culture of Rabbit Primordial Germ Cells

To investigate the characteristics of PGCs and establish rabbit EG (rEG) cells, PGCs were cultured in vitro with various combinations of LIF, bFGF and forskolin on inactivated MEF feeder layers. Kakegawa and colleges (2008) found PGC proliferation in early cultures and induction of EG-like colonies. These cells expressed pluripotent markers, such as alkaline phosphatase activity, Oct-4, Sox-2 and SSEA-1, in the undifferentiated state; however, did not develop into teratoma, when injected into the kidney capsules of SCID mice. The restricted differentiation potential to neural cells were determined via embryoid body formation. Based on those results and the appearance of the germ stem cell markers Vasa, SCP-1 and SCP-3, they suggested that these were hybrid cells with characteristics somewhere between PGC and EG cells.

9.3.3 Differential Expression of Factors Affecting the Maintenance of Pluripotency of ESCs

It is generally accepted, that LIF and Oct-4 play important roles in maintaining mouse ES cell self-renewal (Niwa et al. 2000). Oct-4 is a POU-family transcription factor that is present in the mouse zygote, the inner cell mass, the epiblast and the primordial germ cell (Pesce and Scholer 2001). Oct-4 is also expressed in ES cells, and has an important role in the maintenance of their pluripotency. LIF triggers the activation of Jak tyrosine kinases, results the activation of STAT3 and maintains the pluripotent phenotypes of ES cells (Raz et al. 1999). In murine ES cells, activation of the LIF/gp130/STAT3 signalling pathway efficiently supports ES-cell self-renewal, whereas in primates, including humans and monkeys, ES cells do not respond to LIF signalling in a similar fashion and require feeder cells to maintain their undifferentiated state. Cynomolgus monkey ES cells, like those from mice, do use LIF/gp130/STAT3 signalling, however, as it has been shown, even though activation of STAT3 is induced by LIF, STAT3 signalling is not necessary for ES self-renewal. It was suggested that pathways other than LIF/STAT3 exist that maintain self-renewal of primate ES cells (Sumi et al. 2004).

The comparison between the transcriptomes of mESC and hESC resulted the conclusion that mESCs are substantially different from human ES cell lines. Those recent data underline the morphological and functional differences between the ES cell lines of the two species (Ginis et al. 2004; Laslett et al. 2003; Rao and Zandstra 2005). For instance, hESCs typically grow in tightly adherent, flattened groups rather than in rounded clumps and tolerate physical separation very poorly. Both types of ESCs grow well in presence of MEF. ESCs of both species can be maintained in culture without a feeder layer, but in this condition hESCs require the presence of fibroblast growth factor-2 and activin, whereas mESCs need LIF and BMP4. In contrast, BMP4 induces trophoblast differentiation in hESCs (Brons et al. 2007; Ludwig et al. 2006; Tesar et al. 2007).

In rat ES-like cells, 629 rat transcripts were analysed by Li et al. (2008). The existence of both Bmp and Activin transcripts suggests a unique signal transduction mechanism, several transcripts of the mTOR signalling pathway (Igf1, Figf, Ddit4 and Fyb) were enriched in the undifferentiated rat ES-like cells. On the other hand, the analysis did not detect the upregulation of the JAK/STAT signalling pathway in rat ES-like cells and LIFR, but not LIF was detected in the rat ES-like cells. It is possible that LIF produced by the feeder layer or provided by the culture media activates LIFR in rat ES-like cells to help maintain them in an undifferentiated state.

The currently available data on ESC characteristics established from different ungulate's species, their morphology and properties are definitively more limited than those available for mESC or hESC. Classical hESC and mESC markers such as Oct4, SSEA1, SSEA4 and alkaline phosphatase are indeed expressed by ungulates ICM and embryo-derived cell lines; however, the same genes are also expressed in the trophectoderm and endoderm of their embryos (He et al. 2006; Kirchhof et al. 2000; Talbot et al. 2002; van Eijk et al. 1999). Nanog, another well-characterized ESC marker in human and mouse looks more reliable since it has been shown to be expressed in porcine ESC-like cell lines (Brevini et al. 2007a, b). Moreover nanog was strongly downregulated in caprine trophectoderm, while being strongly expressed in the ICM (He et al. 2006; Keefer et al. 2007). Addition of cytokines and growth factors, such as EGF, LIF, SCF and FGF, that inhibit spontaneous differentiation of hESC and mESC have been reported to be ineffective porcine and bovine epiblast and ICM derived cells (Moore and Piedrahita 1997; Talbot et al. 1995, 1993). The presence of LIF in culture medium seems to inhibit the differentiation process of porcine ESCs because it prevents embryoid body formation (Brevini et al. 2007a). Preliminary information suggests that LIF receptor and gp130 are both expressed in bovine ICM and in its preliminary outgrowth (Pant and Keefer 2006).

To reveal the role of LIF in rabbit ES cell establishment Catunda et al. (2008) isolated and sequenced a partial LIFR cDNA clone. By aligning it with mouse, human and rat sequences a high degree of homology was found, which was underlined by the conservation of several functionally important aminoacid motifs. Rabbit LIFR appears first in 6 dpc rabbit embryos (Fig. 9.4). The rabbit ES like cells were dependent on the presence of LIF in the culture medium and showed marked improvement in their morphology in its presence (Catunda et al. 2008). However contradictory observations were published by Fang et al. (2006): the growth of their rabbit ES cells was dependent on mouse feeder cells, they survived and growed independently of exogenous bFGF and maintained an undifferentiated state in the absence of added LIF. On the other hand, Honda et al. (2008) used 10^3 U/ml ESGRO LIF (Invitrogen) for cultivation of their rabbit ESC lines. Since the expression of CD9 was proved to be associated with LIF-mediated maintenance of undifferentiated murine ES cells (Oka et al. 2002), its detection in rabbit ES cell lines (Fig. 9.4) support that the LIF-LIFR pathway has significant role of in the self-renewal mechanism of rabbit ES cells.

9.3.4 mRNA Expression Pattern of Pluripotency Marker Genes in Rabbit, Human and Mouse ESCs

Wang et al. (2007) analyzed a large set of genes related to pluripotency, including many genes of FGF, TGF-/BMP, and WNT signalling by RT-PCR. The expression pattern in the ESCs of rabbit (rabESC)(RF, RP01, RP2) human (hESC) (BG02) and mouse m(ESC)(R1) were compared (Fig. 9.5). PCR conditions and primers used by them are listed in Table 9.1A. Expression pattern of some genes of the WNT and TGF-/BMP signalling pathways was different in these three species. Inhibitors Dkk2 and Gsk3-ß were not expressed in rabbit ESCs (RF and RP cell lines) but were expressed in human and mouse ESCs, and ligands WNT1, WNT2, WNT4, and WNT5A were expressed in rabbit ESC lines, but only WNT4 was detected in human and mouse ESCs. Nodal, a TGF signalling ligand gene was not detected in rabbit ESCs, but did express in human and mouse ESCs. Primers used to examine LIFR expression are described under Table 9.1B.

9.3.5 Detection of Pluripotency Marker Proteins in Rabbit ES Like Cells with Immunohistochemistry

One of the criteria for pluripotent stem cells is their expression a characteristic set of pluripotency marker genes. Due to the limited availability of antibodies reactive to rabbit antigens, mainly human and mouse specific antibodies were used at immunostaining. In rabbit ES cell lines the classical pluripotency markers were detected independently and partially complementary to each other by

ROLE OF LIFR IN EARLY EMBRYOGENESIS AND STEM CELL ESTABLISHMENT OF RABBITS

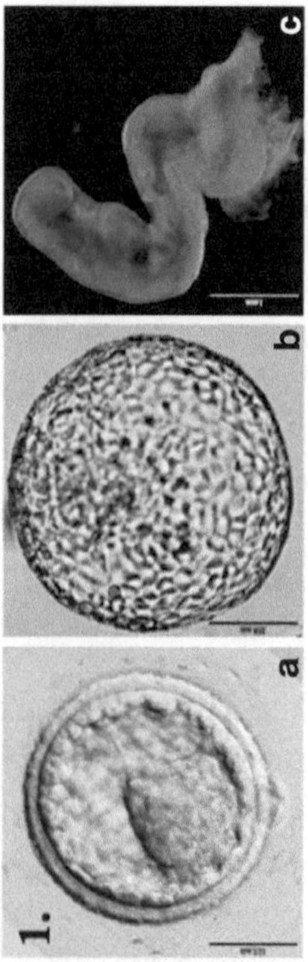

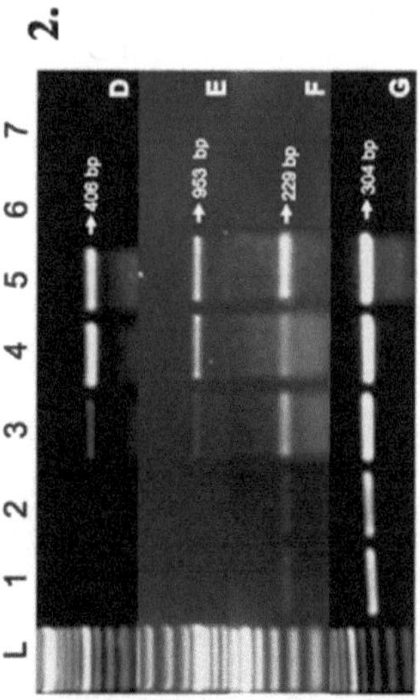

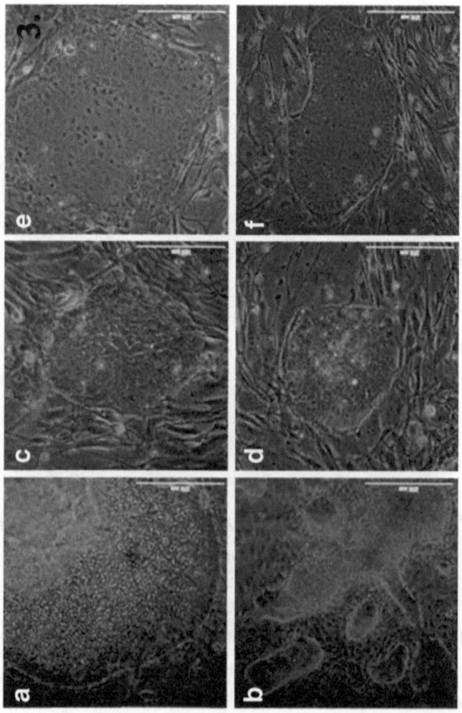

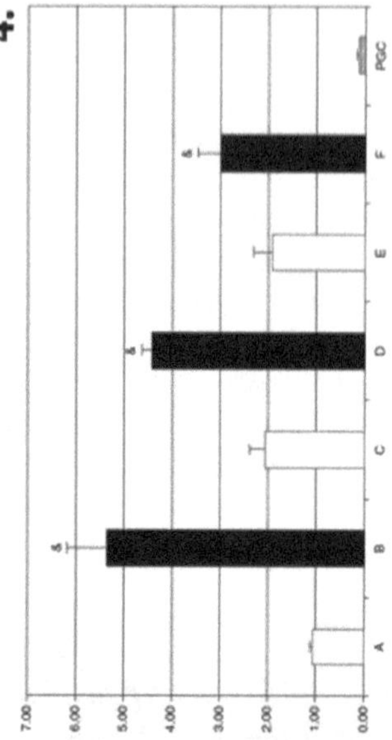

three research groups (Catunda et al. 2008; Honda et al. 2008; Wang et al. 2007). In most cases rabbit ES cells were fixed in 4% paraformaldehyde for 10–20 min at room temperature. SSEA-1, SSEA-3, SSEA-4, tumor-related antigen (TRA)-1–60, TRA-1–81, and Oct-4 (Chemicon International, Temecula, CA, USA) were used by Wang et al. (2007). Honda et al. (2008) used anti-SSEA1, anti-SSEA3, and anti-SSEA4 (Developmental Studies Hybridoma Bank, Iowa, USA), anti-Oct4 from Santa Cruz Biotechnology (Santa Cruz, CA, USA), and anti-Nanog (Cosmobio, Tokyo, Japan). All antibodies were diluted in blocking solution (SSEA1 and Oct4, 1:500; SSEA3, SSEA4, and Nanog, 1:100) and incubated with samples overnight at 4°C. Catunda et al. (2008) analysed the expression patterns of CD9, Nanog, Oct-3/4 and SSEA-1 (Fig. 9.6). The primary antibodies (Human Embryonic Stem Cell Marker Antibody Panel Plus, R&D Systems, USA) were used at a concentration of 1 mg/100 ml. FITC-conjugated anti-mouse IgM and Cy3 conjugated anti-goat IgG and Cy3-conjugated anti-mouse IgG (both from Jackson ImmunoResearch, USA) 1:400 dilution, were used as secondary antibodies.

9.3.6 Genetic Transformation of Rabbit ES Like Cells

Aiming to produce gene-targeted transgenic rabbits as a first step, creation of genetically modified rabbit ESCs were published recently by Fang et al. (2006) and Honda et al. (2008). Fang et al. (2006) used for transformation the pCX-mRFP-neo plasmid, which contains the CMV enhancer, the chicken β-actin promoter, the monomeric red fluorescent protein coding DNA sequences, the rabbit β-globin polyA site and the neo expression cassette. The purified DNA was resuspended in sterile PBS at a concentration of 1 μg/μl. 1.0–1.5 × 10^7 rabbit ES cells were dissociated with 0.05% trypsin/1 mM EDTA resuspended in 800 μl PBS with 100 μg linearized DNA, placed in a 4 mm gap Biorad gene pulser cuvette and set at RT for 20 min. Electroporation was performed at 250 V and 500 μF using the Biorad

Fig. 9.4 **1:** LIFR, gp130, and LIF expression in the differentiating rabbit embryo: (**a**) 3.5 dpc embryo; (**b**) 6 dpc embryo; (**c**) 9 dpc embryo; **2:** (**d**) LIFR expression; (**e**) gp130; (**f**) LIF expression; (**g**) GAPDH expression: L – ladder, 1 – 3.5 dpc blastocyst, 2 – 4.5 dpc blastocyst, 3 – 6 dpc embryo, 4 – 9 dpc rabbit embryo, 5 – 13.5 dpc fetus, 6[mg] negative control (without DNA), 7 – negative control (without transcriptase). Size of bars: A, 125 μm, B, 250 μm, C, 1 mm. **3:** Morphology of rabbit ES-like colonies during rabbit ES cell line derivation. **a**, **c**, **e**: rabbit ES-like cell line derivation without LIF; **b**, **d**, **f**: rabbit ES like cell line derivation with LIF; **a**, **b**: attached ICM; **c**, **d**: ES-like cell colonies after first passage; E,F: ES-like cell colonies after second passage Gp: **4:** Relative expression of LIFR in rabbit ES cell colonies, cultured with or without LIF, as determined by quantitative RT-PCR. **a** – rabbit ICM without LIF, **b** – rabbit ICM with LIF, **c** – first passage of rabbit ES cells without rLIF, **d** – first passage of rabbit ES cells with LIF, **e** – second passage of rabbit ES cells without LIF, **f** – second passage of rabbit ES cells with LIF, PGC – rabbit 13.5 dpc primordial germ cell. Results are presented as the mean SEM. Significant with P 0.05 when comparing samples in medium without and with LIF. Size of bars: 250 μm (Adapted from Catunda et al. (2008) with the permission from Mary Ann Liebert Inc.)

MOLECULAR CHARACTERIZATION OF RABBIT ES-LIKE CELL COLONIES

	rabESC	hESC	mESC
AP	+	+	+
SSEA-1	+	-	+
SSEA-3	+	+	-
SSEA-4	+	+	-
TRA-1-60	+	+	-
TRA-1-81	+	+	-
OCT 4	+	+	+
SOX 2	+	+	+
UTF-1	+	+	+
Nanog	+	+	+
LIFR	+	+,-	+
gp130	+	+	+
LIF	+	+	+
CD9	+	+	+

	rabESC	hESC	mESC
Nodal	-	+	+
TGF-β1	+	+	+
BMP4	+	+	+
WNT1	+	-	-
WNT2	+	-	-
WNT3A	-	-	-
WNT4	+	+	+
WNT5A	+	-	-
DKK1	+	+	+
DKK2	-	+	+
Gsk3-β	-	+	+
FGF1,2	+	+	+
FGFR 1,2,3,4	+	+	+

Fig. 9.5 Genes related to pluripotency, including many genes of FGF, TGF-β/BMP, and WNT signaling, were detected in ESCs of rabbit (rabESC), human (hESC (BG02)), and mouse (mESC (R1)). Some gene expression of WNT and TGF-β/BMP signaling was different in these three species. Inhibitors Dkk2 and Gsk3-β were not expressed in rabbit ESCs (RF and RP cell lines) but were expressed in human and mouse ESCs and ligands WNT1, WNT2, WNT4, and WNT5A were expressed in rabbit ESC lines, but only WNT4 was detected in human and mouse ESCs. Nodal, a TGF signaling ligand gene, was not detected in rabbit ESCs but did express in human and mouse ESCs. *Abbreviations:* +, detected; –, not detected; BMP, bone morphogenetic protein; ESC, embryonic stem cell; FGF, fibroblast growth factor; TGF-β transforming growth factor-β (Adapted from Wang et al. (2007) and Catunda et al. (2008) modified by Gócza and Bősze)

GenePulse II Electroporation System. Cells were then resuspended in the culture medium and plated on new feeders at a density of 1×10^6 cells per 6 cm culture dish. Forty-eight hours post-transfection, 350 μg/ml of G418 was added to the medium. Surviving colonies were selected, mechanically divided into small clumps (about 50 cells) with a glass capillary, and transferred to new feeders in a 3.5 cm culture dish.

Honda et al. (2008) used the self-inactivating lentiviral vector construct pCS CDF-UbC-GFP-PRE, which contains the green fluorescent protein gene under the control of the human ubiquitin C (UbC) promoter. Rabbit ES cells were cultured with lentiviruses at a multiplicity of infection (MOI) of 5, 50, or 200 in flat-bottomed 48-well plates overnight at 37°C, in 6% CO_2 in air. Two days after transduction, ES cells were harvested by trypsinization and were replated onto 12-well plates. Five days after transduction, GFP positive colonies were removed, mechanically dissociated into small clumps, and then cultured in flat-bottomed six-well plates. GFP-positive cells were subsequently propagated under standard conditions (Fig. 9.7).

Table 9.1 Polymerase chain reaction primers and conditions for gene analysis (A: Based on the publication by Wang et al. (2007); B: Based on the publication by Catunda et al. (2008))

A. Genes	Primer sequences	Annealing temperature (°C)	Products (base pairs)	Accession no. or reference
Nanog	5′-AGCAGAAGATGCGGACTG-3′ 5′-GGTCTGGCTGCTCCAAGT-3′	56	350	NM_024865 XM_132755
SOX2	5′-AGCATGATGCAGGAGCAG-3′ 5′-GGAGTGGGAGGAAGAGGT-3′	56	270	NM_003106 NM_011443
UTF-1	5′-GCGCTCACTACTGCTGGAC-3′ 5′-CCCATGAGTTGGCGGATC-3′	58	192	NM_003577 NM_009482
OCT4	5′-GACAACAATGAGAACCTTCA-3′ 5′-CACATCCTTCTCTAGCCCAA-3′	54	185	NM_002701 NM_013633
FGF1	5′-GGAGCGACCAGCACATTC-3′ 5′-TCCCGTTCTTCTTGAGGC-3′	56	237	NM_000800 NM_010197
FGF2	5′-GAAGAGCGACCCTCACATC-3′ 5′-CCCAGTTCGTTTCAGTGCC-3′	59	232	NM_002006 NM_008006
FGFR1	5′-AGATAACACCAAACCAAACCGTATG-3′ 5′-GCATGCAATTTCTTTTCCATCTT-3′	56	330	UniSTS:273319
FGFR2	5′-AAGGTACGAAACCAGCACTGGAG-3′ 5′-TCCATCTCCGTCACATTGAACAG-3′	56	380	UniSTS:464789
FGFR3	5′-CGGAAAGTTCGTCGCTGG-3′ 5′-TTACTGGGCCCTGAGTCTGG-3′	54	150	UniSTS:34762
FGFR4	5′-GGTGCAGACATGAGCAAGG-3′ 5′-CAAGAAGCCGAGCAGAACC-3′	54	270	UniSTS:55376
SOS1	5′-CCACCTCAGGAGAACAAA-3′ 5′-CTCATACGGGTCAAATGC-3′	53	308	NM_005633 NM_009231
PTPN11	5′-AGAGCAACGACGGCAAGT-3′ 5′-TCAAAGGGCAGGATGTTT-3′	53	382	NM_002834 NM_011202

(continued)

Table 9.1 (continued)

A. Genes	Primer sequences	Annealing temperature (°C)	Products (base pairs)	Accession no. or reference
Nodal	5′-GAGCAGAAAAGTGTTGGC-3′ 5′-CTCACAGCGATAGGCATT-3′	53	326	NM_018055 NM_013611
TGF 1	5′-CTGTGGCTACTGGTGCTGAC-3′ 5′-CATTAGCACGCGGGTGACCT-3′	61	294	NM_000660 NM_011577
BMP4	5′-AAAGTCGCCGAGATTCAGG-3′ 5′-GAGTCTGATGGAGGTGAGT-3′	58	634	AY 409113 BC 052846
LeftyA	5′-GATGAAGTGGGCCGAGAA-3′ 5′-CTGGGTGATGGACAAGAAAA-3′	55	379	NM_003240 NM_177099
SMAD1	5′-GTTCAGGCAGTTGCTTACGA-3′ 5′-AGGCATTCCGCATACACC-3′	55	266	NM_001003688 NM_008539
SMAD2	5′-CTTGATGGCCGTCTTCAGGT-3′ 5′-GGGCGGCAGTTCTGTTAG-3′	57	246	NM_001003652 NM_010754
SMAD4	5′-AGGTGGCTGGTCGGAAAG-3′ 5′-TTGGCGGGTGTTGGATGGT-3′	57	312	NM_005359 NM_008540
WNT1	5′-GGGTATTGTGAACGTAGCCTCCTC-3′ 5′-GCCTGCCTCGTTGTTGTGAAG-3′	62	514	NM_005430 NM_021279
WNT2	5′-TCTGGCTCCCTCTGCTCTTGAC-3′ 5′-CAGCCTTCCTGCCAGCTCTGTT-3′	62	565	NM_003391 NM_023653
WNT3A	5′-GTGCTTCTCCACCACCATCTC-3′ 5′-CTTTGTCCACGCCATTGCCTC-3′	62	409	NM_033131 NM_009522
WNT4	5′-GGATGCTCTGACAACATCGCCTA-3′ 5′-CACTTGACGAAGCAGCACCA-3′	60	554	AY419208 AY419210
WNT5A	5′-CATGCAGTACATCGGAGAAGG-3′ 5′-GCAAAGCGGTAGCCATAGTC-3′	60	312	NM_003392 BC018425
DKK1	5′-CCAACGCTATCAAGAACCTGC-3′ 5′-TCCTGAGGCACAGTCTGATG-3′	60	479	NM_012242 NM_010051

DKK2	5′-TCCCAGTCACTGAGAGCA-3′	57	220	NM_014421 NM_020265
	5′-GTCCAGAAGTGGCGAGCA-3′			
Gsk3-	5′-CTACCTTAACCTGGTGCTG-3′	54	270	NM_002093 NM_019827
	5′-TATGAAACATTGGGCTCT-3′			
-Catenin	5′-TGGCAATCAAGAGAGCAA-3′	54	453	NM_001904 NM_007614
	5′-CTGTCACCAGCACGAAGG-3′			
Kremen	5′-TCCAGCATCCATACAACA-3′	52	303	NM_032045 NM_032396
	5′-ACTCCATCCCAGCAAACT-3′			
LRP6	5′-TCCGTCATGCCATTGCCATAGA-3′	60	229	NM_002336 NM_008514
	5′-TCCCATTGAGCCTTGTCACTTC-3′			
PAX6	5′-CATGCAGAACAGTCACAGCGG-3′	60	414	NM_000280 NM_013627
	5′-CCCATCTGTTGCTTTTCGCTA-3′			
AFP	5′-AAATACATCCAGGAGAGCCA-3′	55	416	Lafuste et al. [44]
	5′-CTGAGCTTGGCACAGATCCT-3′			
GAPDH	5′-GGAGCCAAACGGGTCATCATCTC-3′	62	233	Soloviev et al. [45]
	5′-GAGGGGCCATCCACAGTCTTCT-3′			

B.

Primer name	Primer sequence (5′ to 3′ orientation)	Use	Accession no	PCR product size (bp)
Ex19–20f	GAGTGTCTGTGAGGGAAGCAG	LIFR expression analysis	EU441881	406
Ex20r	GTCGTCCTCTACAGCCGTGG			
RabLIFf	TTCTCTATTACACAGCCCAAGG	LIF expression analysis	AX 798851	229
RabLIFr	GTGGCGTTGAGTTTGCTGT			
qLIFRf	TCGGAAGCGAGAATGGATTA	LIFR expression-qPCR	EU441881	104
qLIFRr	CACTGCTTCCCTCACAGACA			
qGAPDHr	GCAGGTCAGGTCCACGAC	GAPDH expression-qPCR	NM001082253	75
RabGAPDHf	GAGCTGAACGGGAAACTCAC	Endogenous control	NM001082253	304
RabGAPDHr	CCCTGTTGCTGTAGCCAAAT			
Rabgp130f	AACGTGGATTCTTCCCACAC	Gp130 expression	ENSOCUT00000 007840	953
Rabgp130r	TCTCCTTGTCCCTCAACACC			

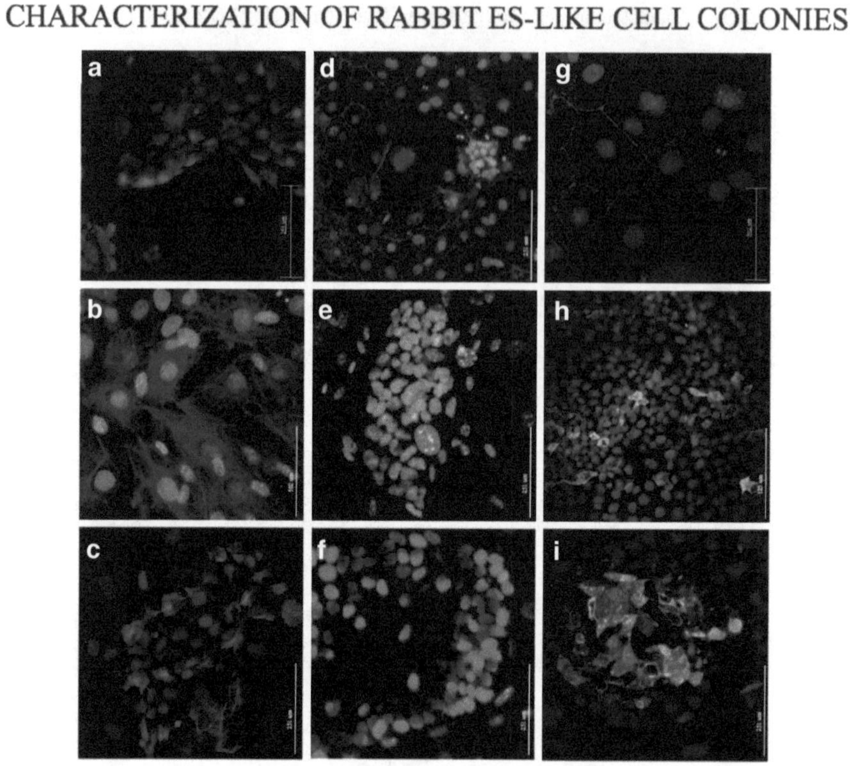

Fig. 9.6 Rabbit ES-like colony morphology showing undifferentiated markers detected by immunohistochemistry. ES-like colonies were labeled with primary antibodies against specific markers and secondary antibodies conjugated with Cy3 (red) or FITC (green) fluorochrome and costained with DAPI (blue). A, B, D, E, G, H: rabbit ES-like cell colonies at third C, F, I: rabbit ES-like cell colonies at fifth passages. Oct4 (E, F), SSEA-1(H, I), Nanog (A, B, C) and CD9 (D, G) expression were detected in the ES-like cell colonies. Size of bars: A, C, D, E, F, I: 251 μm, B, G: 502 μm, H: 125 μm (Adapted from Catunda et al. (2008) with the permission from Mary Ann Liebert Inc.)

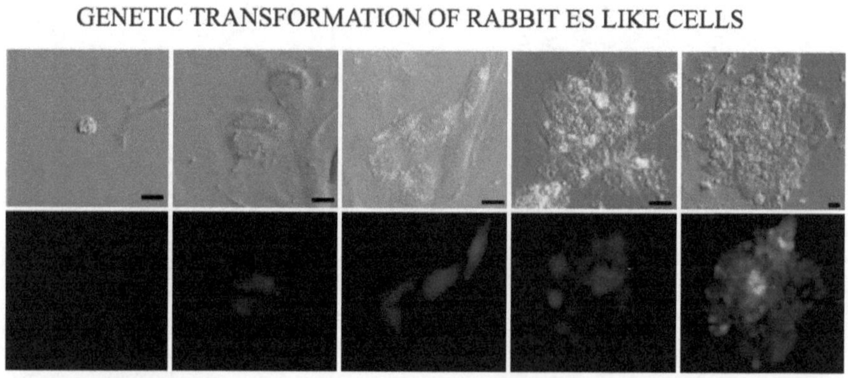

Fig. 9.7 Clonal expansion of rabbit ES cells. Time course of colony formation from a single rabbit ES cell that was transfected with the EGFP gene by a lentiviral vector (at passage 12). The ES cell was seeded onto feeder cells and cell proliferation was scored at 2, 12, 24, 48, and 120 h. (scale bar: 20 μm) (Adapted from Honda et al. (2008) with the permission of Ingenta Connect)

9.3.7 *In Vitro Differentiation of Rabbit ES Cells*

By in vitro cultivation of ES cells as three dimensions aggregates – which are called embryoid bodies (EB) – ES cells can differentiate into derivatives of all three primary germ layers. Upon removing LIF and adding appropriate differentiation agents, ES cells can further differentiate into a variety of cell lineages, including cardiac cells, skeletal-muscle cells, neurons or adipocytes. ES cells can be forced into contact by suspending them in hanging drops of medium. The hanging drop differentiation method has several advantages, including low variation in size of the EBs due to a defined number of ES cells in the starting aggregates, as well as greater reproducibility of differentiation (Wobus and Boheler 2005). The main steps of EB formation in rabbit are presented in Fig. 9.8.

Wang et al. (2007) maintained the rabbit ES cells at high densities for 2–4 weeks without feeder layer replacement. For EB formation, ES cells were digested with 10 mg/ml dispase, resuspended in ESC culture medium, and cultured in hanging drops (30 µl/drop, 40 cells per µl). Two days later, aggregated simple EBs were transferred to bacterial culture dishes (Becton, Dickinson and Company) coated with agar (Sigma-Aldrich) to maintain continuous suspension cultures.

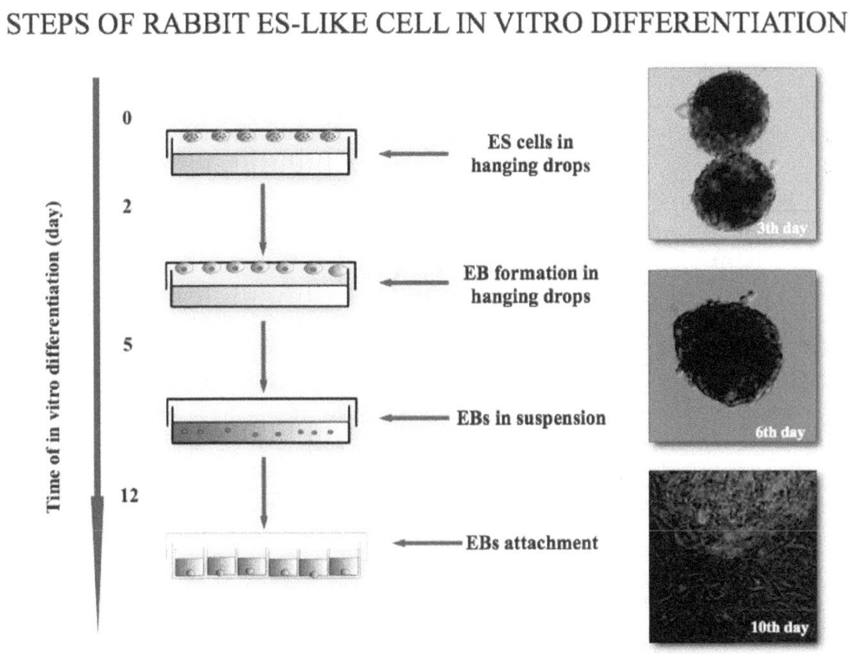

Fig. 9.8 From the rabbit embryonic stem cell colonies growing in ES medium for two passages hanging drops were prepared. In the embryoid bodies placed onto the surface of 0.1% gelatin coated culture dishes fibroblast-like structures were grown out. Contracting cardiac bodies can be identified from the third day until the 12th day (Adapted from Catunda et al. (2008) with permission from Mary Ann Liebert Inc.)

After another 2 days, the resultant cystic EBs were plated into gelatin-coated six-well plates and continuously cultured in ESC culture medium. Cell markers were assessed by immunhistochemistry or RT-PCR. Primary antibodies CK7 (DakoCytomation, Glostrup, Denmark), vimentin (DakoCytomation), α-fetoprotein (AFP) (Sigma-Aldrich), albumin (Sigma-Aldrich), smooth muscle marker actin (DakoCytomation), nestin (Chemicon International), and glial fibrillary acidic protein (GFAP) (Chemicon International) were used as described. After ESCs had differentiated for 3 weeks, specialized cell types representing all three germ cell layers were detected.

For EB formation Honda et al. (2008) digested the rabbit ES cells with 0.05% trypsin, resuspended in a solution containing 78% DMEM/F-12, 20% fetal bovine serum (FBS), 2 mM GlutaMax, 1% nonessential amino acids, and 0.1 mM β-mercaptoethanol and cultured in hanging drops. EBs were collected after 4–7 days in suspension culture and transferred to plastic dishes coated with gelatin to promote adherence. Plating of EBs onto a gelatin-coated culture dish resulted in outgrowths of cells of various types, including multiple cystic structures. In a neural differentiation medium, ES cells differentiated into β-tubulin III-positive neurons and GFAP-positive astrocytes. In the epithelial differentiation medium, they differentiated into MUC1-positive or collagen IV-positive epithelial cells.

Catunda and her coworkers (2008) prepared hanging drops using ES cell culture medium without LIF, containing 500 rabbit ES-like cells in 30 μl (day 0). They kept the embryoid bodies (EBs) in hanging-drop culture for 2 days. On the second day EBs were placed onto the surface of 0.1% gelatin coated culture dishes into ES cell culture medium. Contracting cardiac bodies were observed from the third day until the 12th day, however from the 5th day fibroblast like structure grow out from the attached EBs.

9.3.8 Teratoma Formation as a Marker of the In Vivo Differentiation Potential of Rabbit ESCs

For teratoma formation, either 2–5 × 10^6 rabbit ES cells were injected under the kidney capsule of 5–8 week-old (Honda et al. 2008), or intramuscularly into 6–8-week-old SCID mice (Wang et al. 2007). After 8–14 weeks, teratomas were observed in both cases, dissected and fixed in 4% paraformaldehyde. Paraffin sections were stained with hematoxylin and eosin. Rabbit embryonic stem cells differentiated into tissues representative of all three germ cell layers: hair follicules (ectoderm), skeletal muscle fibers (mesoderm) and glands (endoderm).

9.3.9 Chimera Production with Rabbit ESCs

Chimeric rabbits created earlier were poorly characterized and were produced at low efficiency (Schoonjans et al. 1996). Recently, Bodo et al. (2004) created chimeric rabbits and followed the fate of the microinjected blastomere using the

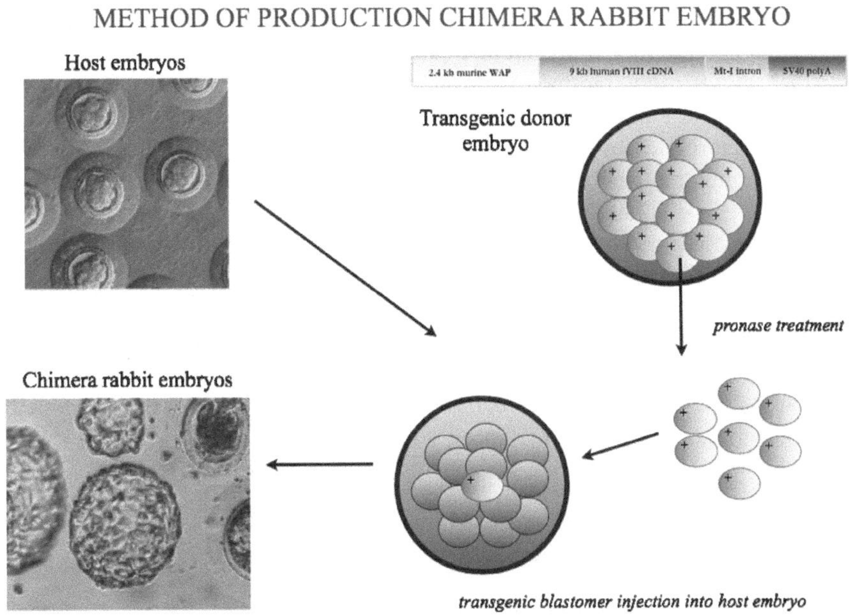

Fig. 9.9 The recipient 16-cell stage embryos were obtained from wild type fertilized does. Five to 20 ES cells derived from transgenic rabbit embryo were injected into the perivitelline spaces of the 16-cell stage embryos by a micro-controller regulated syringe. Following micromanipulation, the chimeric construct embryos were cultured in vitro until compaction in RDH medium. The compacted morula stage embryos were transferred to each oviduct of the recipient does by laparoscopy or cultured in vitro until hatching (Method based on the publication by Bodo et al. (2004))

human factor VIII transgene (Hiripi et al. 2003) as marker (Fig. 9.9). To estimate the level of chimerism of different tissues real-time PCR was used. Eight cell embryos were obtained 44 h after insemination. From the flushed precompacted transgenic embryos the mucin coat and zona pellucida were removed by incubation in 0.5% pronase (Sigma) solution for 10 min. The blastomeres were separated individually by pipetting the zona free embryos up and down in a glass capillary tube several times. The recipient 16-cell stage embryos were obtained from wild type fertilized does. Only one blastomere was injected into the recipient embryo by a microcontroller regulated syringe. Following micromanipulation, the chimeric construct embryos were cultured in vitro until compaction in RDH medium (Jin et al. 2000). Four to twelve compacted morula stage embryos were transferred to each oviduct of the recipient does by laparoscopy (Besenfelder 1998). In their experiments chimeric construct embryos gave rise to chimeric rabbits at 13% efficiency. To estimate the level of chimerism of different tissues real-time PCR was used. The single microinjected blastomere was able to populate all examined tissue types, including the germ line (Bodo et al. 2004).

Honda et al. (2008) stepped forward and tried to produce ES chimeric rabbits with an in vitro transformed transgenic rabbit ES cell line, which is expressing

GFP as marker. Recipient embryos were recovered at the eight-cell or blastocyst stage from females following natural mating. Five to twenty ES cells were injected into the cavities of the blastocysts or into the perivitelline spaces of the eight-cell embryos using a Piezo-driven micromanipulator. The chimerism of the newborn pups was determined by the presence of GFP fluorescence. After the transfer of 47 embryos into recipient females, 17 pups were born in four different experiments using three ES cell lines, however, based on the GFP fluorescence detection the ES cells did not contributed to any kind of tissues.

Lining up with the observations by Honda et al. (2008), it was published recently, that rat ES-like cells expressing a panel of pluripotent cell markers – similar to those described in rabbit ES cells – were able to contribute to the developing extraembryonic, but not to embryonic tissues (Demers et al. 2007). It remains to be answered, whether the currently available rabbit ES cells would be able to contribute to the extraembryonic lineages too.

9.3.10 Nuclear Transplantation Using Rabbit ES Like Cells

Nuclear transplantation has been envisaged as a mean to produce cells or tissues for human autologous transplantation. Understanding the factors affecting the efficiency of rabbit ES like cell nuclear transplantation, could be a good model for human nuclear transplantation. There are reports about successful nuclear transplantation using rabbit ES like cells. Rabbit embryonic stem-like cells, characterized by embryoid body formation and differentiation into cell types representative of all three germ layers, were studied for their ability to promote early embryonic development after nuclear transfer (Du et al. 1995). The cloned embryos from the embryonic stem-like cells appeared normal, with an average of 26% inner cell mass cells, similar to that of control non-manipulated embryos (25%) or cloned embryos from blastomeres (25%).

Chen and collegues (2003) reported about embryonic stem cell line derivation by the transfer of human somatic nuclei into rabbit oocytes. Those results suggest, that human somatic nuclei can form ntES cells, independent of the age of the donor. The ntES cells maintain the capability of sustained growth in an undifferentiated state, and form embryoid bodies, which give rise to cell types such as neuron and muscle, as well as mixed cell populations that express markers representative of all three germ layers.

Embryonic stem cells were isolated from rabbit blastocysts derived from fertilization, parthenogenesis and nuclear transfer and propagated in a serum-free culture system (Fang et al. 2006). The rabbit ES cells derived from nuclear transfer proliferated for a prolonged time in an undifferentiated state and maintained a normal karyotype, formed embryoid bodies and generated teratoma that contained tissue types of all three germ layers. Notably, those ES cells exhibited a high cloning efficiency, were genetically modified readily and were used as nuclear donors to generate a viable rabbit through somatic cell nuclear transfer.

9.4 Discussion

Stable animal ES cell lines have only been generated from mice and monkeys (Thomson et al. 1995). Extrapolation of results obtained using mice to humans is limited due to fundamental differences between murine and human ES cells (Koestenbauer et al. 2006). The use of monkey cell lines may hinder progress because of the limitations of transplantation trials. Human ES cells are potentially useful for regenerative medicine, developmental biology, tissue regeneration, disease pathology, and drug discovery (Thomson et al. 1998). The use of human ES cell lines is also limited, because destruction of developing human embryos is required for their establishment (Lerou et al. 2008). Recently, Wang et al. (2007) and Honda and coworkers (2008) described novel methods for the establishment of rabbit ES cell lines with indefinite proliferative ability. Derivation of those ES cell lines were achieved primarily by optimization of feeder cell conditions and secondarily by an improved zona-shedding method for blastocysts. Further experiments are necessary, to prove the usefullness of this technology. Potentially, an efficient way of rabbit ES cell isolation could be a viable alternative as a small animal model of human ES cells.

In addition to their usefulness as a regeneration therapy model, ES cells have great potential for the production of genetically modified animals, especially gene targeted animals which so far restricted to mouse. Potentially, the rabbit have mutliple advantages over mice in size and physiology, therefore transgenic rabbit models based on targeted genetic alterations could be the appropriate models of human diseases in preclinical trials. Since germline transmission of rabbit ESCs so far was unsuccessful, the combination of rabbit ESCs with genetic alteration and animal cloning technologies could be a viable alternative.

Acknowledgements This work was supported by grants OM-00367/2004, OTKA T037582, OTKA T049034, OTKA K77913, GVOP-0071/3.0 and MTA-JSPS/102.

References

Amit M., Carpenter M. K., Inokuma M. S., Chiu C. P., Harris C. P., Waknitz M. A., Itskovitz-Eldor J. and Thomson J. A. (2000) Clonally derived human embryonic stem cell lines maintain pluripotency and proliferative potential for prolonged periods of culture. *Dev Biol* **227**, 271–278.

Besenfelder U. (1998) Generation and application of transgenic rabbits. In *Microinjection and transgenesis: strategies and protocol*, pp. 561–586. Springer Verlag, Berlin.

Bodo S., Gocza E., Revay T., Hiripi L., Carstea B., Kovacs A., Bodrogi L. and Bosze Z. (2004) Production of transgenic chimeric rabbits and transmission of the transgene through the germline. *Mol Reprod Dev* **68**, 435–440.

Bosze Z. and Houdebine L. -M. (2006) Application of rabbits in biomedical research: a review. *World Rabbit Sci* **14**, 1–14.

Bradley A., Evans M., Kaufman M. H. and Robertson E. (1984) Formation of germ-line chimaeras from embryo-derived teratocarcinoma cell lines. *Nature* **309**, 255–256.

Brandenberger R., Wei H., Zhang S., Lei S., Murage J., Fisk G. J., Li Y., Xu C., Fang R., Guegler K., Rao M. S., Mandalam R., Lebkowski J. and Stanton L. W. (2004) Transcriptome characterization elucidates signaling networks that control human ES cell growth and differentiation. *Nat Biotechnol* **22**, 707–716.

Brevini T. A., Antonini S., Cillo F., Crestan M. and Gandolfi F. (2007a) Porcine embryonic stem cells: facts, challenges and hopes. *Theriogenology* **68 Suppl 1**, S206–213.

Brevini T. A., Tosetti V., Crestan M., Antonini S. and Gandolfi F. (2007b) Derivation and characterization of pluripotent cell lines from pig embryos of different origins. *Theriogenology* **67**, 54–63.

Brevini T., Antonini S., Pennerossa G. and Gandolfi F. (2008) Recent progress in embryonic stem cell research and its application in domestic species. *Reprod Domest Anim* **43**, 193–199.

Brinster R. L. (1974) The effect of cells transferred into the mouse blastocyst on subsequent development. *J Exp Med* **140**, 1049–1056.

Brons I. G., Smithers L. E., Trotter M. W., Rugg-Gunn P., Sun B., Chuva de Sousa Lopes S. M., Howlett S. K., Clarkson A., Ahrlund-Richter L., Pedersen R. A. and Vallier L. (2007) Derivation of pluripotent epiblast stem cells from mammalian embryos. *Nature* **448**, 191–195.

Catunda A. P., Gocza E., Carstea B. V., Hiripi L., Hayes H., Rogel-Gaillard C., Bertaud M. and Bosze Z. (2008) Characterization, Chromosomal Assignment, and Role of LIFR in Early Embryogenesis and Stem Cell Establishment of Rabbits. *Cloning Stem Cells* **10**, 523–534.

Chen Y., He Z. X., Liu A., Wang K., Mao W. W., Chu J. X., Lu Y., Fang Z. F., Shi Y. T., Yang Q. Z., Chen da Y., Wang M. K., Li J. S., Huang S. L., Kong X. Y., Shi Y. Z., Wang Z. Q., Xia J. H., Long Z. G., Xue Z. G., Ding W. X. and Sheng H. Z. (2003) Embryonic stem cells generated by nuclear transfer of human somatic nuclei into rabbit oocytes. *Cell Res* **13**, 251–263.

Choo A., Padmanabhan J., Chin A., Fong W. J. and Oh S. K. (2006) Immortalized feeders for the scale-up of human embryonic stem cells in feeder and feeder-free conditions. *J Biotechnol* **122**, 130–141.

Cole R. J., Edwards R. G., and Paul J. (1966) Cytodifferentiation and embryogenesis in cell colonies and tissue cultures derived from ova and blastocyst of the rabbit. *Dev Biol* **13**, 385–407.

Daheron L., Opitz S. L., Zaehres H., Lensch W. M., Andrews P. W., Itskovitz-Eldor J. and Daley G. Q. (2004) LIF/STAT3 signaling fails to maintain self-renewal of human embryonic stem cells. *Stem Cells* **22**, 770–778.

Demers S. P., Yoo J. G., Lian L., Therrien J. and Smith L. C. (2007) Rat embryonic stem-like (ES-like) cells can contribute to extraembryonic tissues in vivo. *Cloning Stem Cells* **9**, 512–522.

Doetschman T., Gregg R. G., Maeda N., Hooper M. L., Melton D. W., Thompson S. and Smithies O. (1987) Targetted correction of a mutant HPRT gene in mouse embryonic stem cells. *Nature* **330**, 576–578.

Du F., Giles J. R., Foote R. H., Graves K. H., Yang X. and Moreadith R. W. (1995) Nuclear transfer of putative rabbit embryonic stem cells leads to normal blastocyst development. *J Reprod Fertil* **104**, 219–223.

Evans M. J. and Kaufman M. H. (1981) Establishment in culture of pluripotential cells from mouse embryos. *Nature* **292**, 154–156.

Fang Z. F., Gai H., Huang Y. Z., Li S. G., Chen X. J., Shi J. J., Wu L., Liu A., Xu P. and Sheng H. Z. (2006) Rabbit embryonic stem cell lines derived from fertilized, parthenogenetic or somatic cell nuclear transfer embryos. *Exp Cell Res* **312**, 3669–3682.

Ginis I., Luo Y., Miura T., Thies S., Brandenberger R., Gerecht-Nir S., Amit M., Hoke A., Carpenter M. K., Itskovitz-Eldor J. and Rao M. S. (2004) Differences between human and mouse embryonic stem cells. *Dev Biol* **269**, 360–380.

Graves K. H. and Moreadith R. W. (1993) Derivation and characterization of putative pluripotential embryonic stem cells from preimplantation rabbit embryos. *Mol Reprod Dev* **36**, 424–433.

Guan K., Nayernia K., Maier L. S., Wagner S., Dressel R., Lee J. H., Nolte J., Wolf F., Li M., Engel W. and Hasenfuss G. (2006) Pluripotency of spermatogonial stem cells from adult mouse testis. *Nature* **440**, 1199–1203.

He S., Pant D., Schiffmacher A., Bischoff S., Melican D., Gavin W. and Keefer C. (2006) Developmental expression of pluripotency determining factors in caprine embryos: novel pattern of NANOG protein localization in the nucleolus. *Mol Reprod Dev* **73**, 1512–1522.

Hiripi L., Makovics F., Halter R., Baranyi M., Paul D., Carnwath J. W., Bosze Z. and Niemann H. (2003) Expression of active human blood clotting factor VIII in mammary gland of transgenic rabbits. *DNA Cell Biol* **22**, 41–45.

Hogan B., Fellous M., Avner P. and Jacob F. (1977) Isolation of a human teratoma cell line which expresses F9 antigen. *Nature* **270**, 515–518.

Honda A., Hirose M., Inoue K., Ogonuki N., Miki H., Shimozawa N., Hatori M., Shimizu N., Murata T., Hirose M., Katayama K., Wakisaka N., Miyoshi H., Yokoyama K. K., Sankai T. and Ogura A. (2008) Stable embryonic stem cell lines in rabbits: potential small animal models for human research. *Reprod Biomed Online* **17**, 706–715.

Humphrey R. K., Beattie G. M., Lopez A. D., Bucay N., King C. C., Firpo M. T., Rose-John S. and Hayek A. (2004) Maintenance of pluripotency in human embryonic stem cells is STAT3 independent. *Stem Cells* **22**, 522–530.

Jin D. I., Kim D. K., Im K. S. and Choi W. S. (2000) Successful pregnancy after transfer of rabbit blastocysts grown in vitro from single-cell zygotes. *Theriogenology* **54**, 1109–1116.

Kahan B. W. and Ephrussi B. (1970) Developmental potentialities of clonal in vitro cultures of mouse testicular teratoma. *J Natl Cancer Inst* **44**, 1015–1036.

Kakegawa R., Teramura T., Takehara T., Anzai M., Mitani T., Matsumoto K., Saeki K., Sagawa N., Fukuda K. and Hosoi Y. (2008) Isolation and culture of rabbit primordial germ cells. *J Reprod Dev* **54**, 352–357.

Kanatsu-Shinohara M., Inoue K., Lee J., Yoshimoto M., Ogonuki N., Miki H., Baba S., Kato T., Kazuki Y., Toyokuni S., Toyoshima M., Niwa O., Oshimura M., Heike T., Nakahata T., Ishino F., Ogura A. and Shinohara T. (2004) Generation of pluripotent stem cells from neonatal mouse testis. *Cell* **119**, 1001–1012.

Keefer C. L., Pant D., Blomberg L. and Talbot N. C. (2007) Challenges and prospects for the establishment of embryonic stem cell lines of domesticated ungulates. *Anim Reprod Sci* **98**, 147–168.

Kirchhof N., Carnwath J. W., Lemme E., Anastassiadis K., Scholer H. and Niemann H. (2000) Expression pattern of Oct-4 in preimplantation embryos of different species. *Biol Reprod* **63**, 1698–1705.

Kleinsmith L. J. and Pierce G. B., Jr. (1964) Multipotentiality of single embryonal carcinoma cells. *Cancer Res* **24**, 1544–1551.

Koestenbauer S., Zech N. H., Juch H., Vanderzwalmen P., Schoonjans L. and Dohr G. (2006) Embryonic stem cells: similarities and differences between human and murine embryonic stem cells. *Am J Reprod Immunol* **55**, 169–180.

Labosky P. A., Barlow D. P. and Hogan B. L. (1994) Mouse embryonic germ (EG) cell lines: transmission through the germline and differences in the methylation imprint of insulin-like growth factor 2 receptor (Igf2r) gene compared with embryonic stem (ES) cell lines. *Development* **120**, 3197–3204.

Laslett A. L., Filipczyk A. A. and Pera M. F. (2003) Characterization and culture of human embryonic stem cells. *Trends Cardiovasc Med* **13**, 295–301.

Lerou P. H., Yabuuchi A., Huo H., Takeuchi A., Shea J., Cimini T., Ince T. A., Ginsburg E., Racowsky C. and Daley G. Q. (2008) Human embryonic stem cell derivation from poor-quality embryos. *Nat Biotechnol* **26**, 212–214.

Levenstein M. E., Ludwig T. E., Xu R. H., Llanas R. A., VanDenHeuvel-Kramer K., Manning D. and Thomson J. A. (2006) Basic fibroblast growth factor support of human embryonic stem cell self-renewal. *Stem Cells* **24**, 568–574.

Li C., Yang Y., Gu J., Ma Y. and Jin Y. (2008) Derivation and transcriptional profiling analysis of pluripotent stem cell lines from rat blastocysts. *Cell Res* **19**, 173–186.

Li T., Wang S., Xie Y., Lu Y., Zhang X., Wang L., Yang S., Wolf D., Zhou Q. and Ji W. (2005) Homologous feeder cells support undifferentiated growth and pluripotency in monkey embryonic stem cells. *Stem Cells* **23**, 1192–1199.

Lowry N., Goderie S. K., Adamo M., Lederman P., Charniga C., Gill J., Silver J. and Temple S. (2008) Multipotent embryonic spinal cord stem cells expanded by endothelial factors and Shh/RA promote functional recovery after spinal cord injury. *Exp Neurol* **209**, 510–522.
Ludwig T. E., Levenstein M. E., Jones J. M., Berggren W. T., Mitchen E. R., Frane J. L., Crandall L. J., Daigh C. A., Conard K. R., Piekarczyk M. S., Llanas R. A. and Thomson J. A. (2006) Derivation of human embryonic stem cells in defined conditions. *Nat Biotechnol* **24**, 185–187.
Maherali N., Sridharan R., Xie W., Utikal J., Eminli S., Arnold K., Stadtfeld M., Yachechko R., Tchieu J., Jaenisch R., Plath K. and Hochedlinger K. (2007) Directly reprogrammed fibroblasts show global epigenetic remodeling and widespread tissue contribution. *Cell Stem Cell* **1**, 55–70.
Martin G. R. (1981) Isolation of a pluripotent cell line from early mouse embryos cultured in medium conditioned by teratocarcinoma stem cells. *Proc Natl Acad Sci U S A* **78**, 7634–7638.
Matsui Y., Zsebo K. and Hogan B. L. (1992) Derivation of pluripotential embryonic stem cells from murine primordial germ cells in culture. *Cell* **70**, 841–847.
Moore K. and Piedrahita J. A. (1997) The effects of human leukemia inhibitory factor (hLIF) and culture medium on in vitro differentiation of cultured porcine inner cell mass (pICM). *In Vitro Cell Dev Biol Anim* **33**, 62–71.
Niwa H., Miyazaki J. and Smith A. G. (2000) Quantitative expression of Oct-3/4 defines differentiation, dedifferentiation or self-renewal of ES cells. *Nat Genet* **24**, 372–376.
Notarianni E., Galli C., Laurie S., Moor R. M. and Evans M. J. (1991) Derivation of pluripotent, embryonic cell lines from the pig and sheep. *J Reprod Fertil Suppl* **43**, 255–260.
Oka M., Tagoku K., Russell T. L., Nakano Y., Hamazaki T., Meyer E. M., Yokota T. and Terada N. (2002) CD9 is associated with leukemia inhibitory factor-mediated maintenance of embryonic stem cells. *Mol Biol Cell* **13**, 1274–1281.
Okita K., Ichisaka T. and Yamanaka S. (2007) Generation of germline-competent induced pluripotent stem cells. *Nature* **448**, 313–317.
Pant D. and Keefer C. (2006) Gene expression in cultures of inner cell masses isolated from in vitro produced and in vivo derived bovine blastocysts. *Reprod Fertil Dev* **18**, 110–110.
Park I. H., Arora N., Huo H., Maherali N., Ahfeldt T., Shimamura A., Lensch M. W., Cowan C., Hochedlinger K. and Daley G. Q. (2008) Disease-specific induced pluripotent stem cells. *Cell* **134**, 877–886.
Pesce M. and Scholer H. R. (2001) Oct-4: gatekeeper in the beginnings of mammalian development. *Stem Cells* **19**, 271–278.
Piedrahita J. A., Anderson G. B. and Bondurant R. H. (1990a) Influence of feeder layer type on the efficiency of isolation of porcine embryo-derived cell lines. *Theriogenology* **34**, 865–877.
Piedrahita J. A., Anderson G. B. and Bondurant R. H. (1990b) On the isolation of embryonic stem cells: comparative behavior of murine, porcine and ovine embryos. *Theriogenology* **34**, 879–901.
Rao B. M. and Zandstra P. W. (2005) Culture development for human embryonic stem cell propagation: molecular aspects and challenges. *Curr Opin Biotechnol* **16**, 568–576.
Raz R., Lee C. K., Cannizzaro L. A., d'Eustachio P. and Levy D. E. (1999) Essential role of STAT3 for embryonic stem cell pluripotency. *Proc Natl Acad Sci U S A* **96**, 2846–2851.
Resnick J. L., Bixler L. S., Cheng L. and Donovan P. J. (1992) Long-term proliferation of mouse primordial germ cells in culture. *Nature* **359**, 550–551.
Rossant J. (2008) Stem cells and early lineage development. *Cell* **132**, 527–531.
Schoonjans L., Albright G. M., Li J. L., Collen D. and Moreadith R. W. (1996) Pluripotential rabbit embryonic stem (ES) cells are capable of forming overt coat color chimeras following injection into blastocysts. *Mol Reprod Dev* **45**, 439–443.
Shamblott M. J., Axelman J., Wang S., Bugg E. M., Littlefield J. W., Donovan P. J., Blumenthal P. D., Huggins G. R. and Gearhart J. D. (1998) Derivation of pluripotent stem cells from cultured human primordial germ cells. *Proc Natl Acad Sci U S A* **95**, 13726–13731.

Shiomi M. and Fan J. (2008) Unstable coronary plaques and cardiac events in myocardial infarction-prone Watanabe heritable hyperlipidemic rabbits: questions and quandaries. *Curr Opin Lipidol* **19**, 631–636.

Stacey G. N., Cobo F., Nieto A., Talavera P., Healy L. and Concha A. (2006) The development of 'feeder' cells for the preparation of clinical grade hES cell lines: challenges and solutions. *J Biotechnol* **125**, 583–588.

Steer H. W. (1970) The trophoblastic knobs of the preimplanted rabbit blastocyst: a light and electron microscopic study. *J Anat* **107**, 315–325.

Stewart C. L. (1994) Leukaemia inhibitory factor and the regulation of pre-implantation development of the mammalian embryo. *Mol Reprod Dev* **39**, 233–238.

Sumi T., Fujimoto Y., Nakatsuji N. and Suemori H. (2004) STAT3 is dispensable for maintenance of self-renewal in nonhuman primate embryonic stem cells. *Stem Cells* **22**, 861–872.

Takahashi K. and Yamanaka S. (2006) Induction of pluripotent stem cells from mouse embryonic and adult fibroblast cultures by defined factors. *Cell* **126**, 663–676.

Talbot N. C., Rexroad C. E., Jr., Pursel V. G., Powell A. M. and Nel N. D. (1993) Culturing the epiblast cells of the pig blastocyst. *In Vitro Cell Dev Biol Anim* **29A**, 543–554.

Talbot N. C., Powell A. M. and Rexroad C. E., Jr. (1995) In vitro pluripotency of epiblasts derived from bovine blastocysts. *Mol Reprod Dev* **42**, 35–52.

Talbot N. C., Powell A. M. and Garrett W. M. (2002) Spontaneous differentiation of porcine and bovine embryonic stem cells (epiblast) into astrocytes or neurons. *In Vitro Cell Dev Biol Anim* **38**, 191–197.

Tesar P. J., Chenoweth J. G., Brook F. A., Davies T. J., Evans E. P., Mack D. L., Gardner R. L. and McKay R. D. (2007) New cell lines from mouse epiblast share defining features with human embryonic stem cells. *Nature* **448**, 196–199.

Thomas K. R. and Capecchi M. R. (1987) Site-directed mutagenesis by gene targeting in mouse embryo-derived stem cells. *Cell* **51**, 503–512.

Thomson J. A., Kalishman J., Golos T. G., Durning M., Harris C. P., Becker R. A. and Hearn J. P. (1995) Isolation of a primate embryonic stem cell line. *Proc Natl Acad Sci U S A* **92**, 7844–7848.

Thomson J. A., Kalishman J., Golos T. G., Durning M., Harris C. P. and Hearn J. P. (1996) Pluripotent cell lines derived from common marmoset (Callithrix jacchus) blastocysts. *Biol Reprod* **55**, 254–259.

Thomson J. A., Itskovitz-Eldor J., Shapiro S. S., Waknitz M. A., Swiergiel J. J., Marshall V. S. and Jones J. M. (1998) Embryonic stem cell lines derived from human blastocysts. *Science* **282**, 1145–1147.

Turnpenny L., Brickwood S., Spalluto C. M., Piper K., Cameron I. T., Wilson D. I. and Hanley N. A. (2003) Derivation of human embryonic germ cells: an alternative source of pluripotent stem cells. *Stem Cells* **21**, 598–609.

van de Lavoir M. C., Diamond J. H., Leighton P. A., Mather-Love C., Heyer B. S., Bradshaw R., Kerchner A., Hooi L. T., Gessaro T. M., Swanberg S. E., Delany M. E. and Etches R. J. (2006) Germline transmission of genetically modified primordial germ cells. *Nature* **441**, 766–769.

van Eijk M. J., van Rooijen M. A., Modina S., Scesi L., Folkers G., van Tol H. T., Bevers M. M., Fisher S. R., Lewin H. A., Rakacolli D., Galli C., de Vaureix C., Trounson A. O., Mummery C. L. and Gandolfi F. (1999) Molecular cloning, genetic mapping, and developmental expression of bovine POU5F1. *Biol Reprod* **60**, 1093–1103.

Vassilieva S., Guan K., Pich U. and Wobus A. (2000) Establishment of SSEA-1- and Oct-4-expressing rat embryonic stem-like cell lines and effects of cytokines of the IL-6 family on clonal growth. *Exp Cell Res* **258**, 361–373.

Wang S., Tang X., Niu Y., Chen H., Li B., Li T., Zhang X., Hu Z., Zhou Q. and Ji W. (2007) Generation and characterization of rabbit embryonic stem cells. *Stem Cells* **25**, 481–489.

Wernig M., Meissner A., Foreman R., Brambrink T., Ku M., Hochedlinger K., Bernstein B. E. and Jaenisch R. (2007) In vitro reprogramming of fibroblasts into a pluripotent ES-cell-like state. *Nature* **448**, 318–324.

Wheeler M. (1994) Development and validation of swine embryonic stem cells: a review. *Reprod Fertil Dev* **6**, 563–568.

Wobus A. M. and Boheler K. R. (2005) Embryonic stem cells: prospects for developmental biology and cell therapy. *Physiol Rev* **85**, 635–678.

Xu R. H., Peck R. M., Li D. S., Feng X., Ludwig T. and Thomson J. A. (2005) Basic FGF and suppression of BMP signaling sustain undifferentiated proliferation of human ES cells. *Nat Meth* **2**, 185–190.

Yamagata K., Nakanishi T., Ikawa M., Yamaguchi R., Moss S. B. and Okabe M. (2002) Sperm from the calmegin-deficient mouse have normal abilities for binding and fusion to the egg plasma membrane. *Dev Biol* **250**, 348–357.

Yu J. and Thomson J. A. (2008) Pluripotent stem cell lines. *Genes Dev* **22**, 1987–1997.

Yu J., Vodyanik M. A., He P., Slukvin, II and Thomson J. A. (2006) Human embryonic stem cells reprogram myeloid precursors following cell-cell fusion. *Stem Cells* **24**, 168–176.

Yu J., Vodyanik M. A., Smuga-Otto K., Antosiewicz-Bourget J., Frane J. L., Tian S., Nie J., Jonsdottir G. A., Ruotti V., Stewart R., Slukvin, II and Thomson J. A. (2007) Induced pluripotent stem cell lines derived from human somatic cells. *Science* **318**, 1917–1920.

Chapter 10
Rabbit Cloning

Andras Dinnyes, Zsuzsanna Polgar, and Qinggang Meng

Abstract Rabbit somatic cell nuclear transfer (SCNT) offers new opportunities for basic research, biomedical and agricultural applications and genome preservation. However, the technology is still in a very early development stage and it is relatively inefficient proving many early expectations premature. The technical steps of SCNT are extremely complex. They are also very sensitive to small changes and fluctuations in technical parameters. Furthermore, the biological background of the reprogramming process is not fully understood. It makes it virtually impossible to optimize a protocol providing "ideal" recipient oocytes and donor cells for the process. In this chapter we will intend to summarize the current status of nuclear transfer technology in rabbits and to give the readers a brief insight of its future use.

Keywords nuclear transfer, cloning, micromanipulation, epigenetics, parthenogenetic activation

10.1 Historical Background of Rabbit Biotechnology

The first successful rabbit embryo transfer was reported in 1890 by Heape. Over the years, rabbit continued to be an important model species for research in reproductive biology. Its oviduct and uterus provided an easy access and low cost

A. Dinnyes (✉)
Molecular Animal Biotechnology Laboratory, Szent Istvan University, Pater K. u. 1., Godollo, H-2103, Hungary; Genetic Reprogramming Group, Agricultural Biotechnology Center, Godollo, Hungary
e-mail: andrasdinnyes@yahoo.com

Z. Polgar
Genetic Reprogramming Group, Agricultural Biotechnology Center, Godollo, Hungary; Constantine the Philosopher University, Faculty of Natural Sciences, Slovakia

Q. Meng
Genetic Reprogramming Group, Agricultural Biotechnology Center, Godollo, Hungary; Department of Animal, Dairy and Veterinary Sciences, Utah State University, Logan, USA

L.-M. Houdebine and J. Fan (eds.), *Rabbit Biotechnology: Rabbit Genomics, Transgenesis, Cloning and Models*,

environment for embryo culture until *in vitro* culture technologies were sufficiently advanced (Chang et al. 1971; Ellington et al. 1990; Totey et al. 1992; Petters and Wells 1993). Rabbits have also been good models for the development and improvement of many micromanipulation techniques. These protocols include the production of transgenic rabbits applying pronuclear microinjection (Hammer et al. 1985), identical twin rabbits by embryo splitting (Yang and Foote 1987), live young from intracytoplasmic sperm injection (Deng and Yang 2001), chimera production by injection of inner cell mass (ICM) cells into morulae or blastocysts (Giles et al. 1993), nuclear transfer with embryonic blastomeres (Stice and Robl 1988) with cumulus (Chesné et al. 2002) and fibroblast (Li et al. 2006) cells and embryonic stem cell line establishment from SCNT embryos (Fang et al. 2006).

10.2 Special Aspects of Rabbit Reproduction

Rabbit embryonal and fetal development provides an important model for other species and has been studied extensively. The main events of the preimplantation development of rabbit embryos are described in Table 10.1. Reference work of embryonic and fetal development and further details of the laboratory rabbits biology can be found in the work of Adams et al. (1961), Edwards (1968), Hagen (1974), Anderson and Henck (1994), and Harkness and Wagner (1995).

Rabbits are classified as induced ovulators because mature preovulatory follicles are present constantly on the surface of their ovaries. A surge in the concentration

Table 10.1 Pre-implantation *in vivo* and *in vitro* embryo development in rabbit (Modified from Dinnyes et al., 2002)

Time post-mating (h)		Embryo-stage (cell number	Doubling time	Location of Embryos	
In vivo	*In vitro*	*in vivo*)	*in vivo* (h)	*in vivo*	Major events
12–14	–	Oocyte	–	Oviduct	Fertilization
18–20	–	Zygote	–	Oviduct	Pronuclear formation (6 h)
24–26	25–27	2-cell	6	Oviduct	–
30–32	32–36	4-cell	6	Oviduct	–
38–40	41–47	8-cell	8	Oviduct	Maternal-zygotic transition
46–48	48–50	16-cell	8	Oviduct	
54–56	62–64	Morula (32-cell)	8	Oviduct	Compaction and transport to uterus
64–66	72–78	Compact morula (64-cell)	10	Oviduct/ uterus	Morula-blastocyst transition
76–78	80–91	Early blastocyst (128-cell)	12	Oviduct/ uterus	
84–86	96–98	Expanded blastocyst (256-cell)	8	Uterus	Blastocoele expansion
94–96	100–110	Hatched blastocyst (512-cell)	8	Uterus	Hatching

of luteinizing hormone (LH) in the blood plasma can induce ovulation, resulting either from the administration of LH or human chorionic gonadotropin (hCG), stimulated by gonadotropin-releasing hormone (GnRH).

Copulation initiates a sequence of neuronal, neuroendocrine and endocrine events that culminate in the LH surge. Initially, the mating provides sensory inputs that are relayed by neuronal pathways to the brain. Neurotransmitter norepinephrine, (NE) is stimulated into the hypothalamus of the brain by neurons whose cell bodies are located in the brainstem. Genome studies indicate that genes involved in NE synthesis exhibit increased expression in the brainstem neurons within 20–40 min after mating. Neurosecretary cells in the hypothalamus initiate a surge in secretion of gonadotropin-releasing hormone, (GnRH), into the blood in the hypothalamo-hypophysial portal system within an hour after copulation.

Mating is the natural mechanism for inducing ovulation. It stimulates mechanically the central nervous system to induce both, the GnRH and LH surges, making the timing of ovulation and, also making the age of the oocytes highly predictable and controllable (Yang 1991). Ovulation normally occurs 10–12 h after the onset of hormonal administration or mechanical stimulus, although superovulation can cause slightly prolonged ovulation periods (Varian et al. 1967).

The mucin coat, a unique glycoprotein covering the rabbit embryos is a characteristic feature of rabbit embryology. It is accumulated during oviductal transport to reach a thickness up to 110 μm 72 h after ovulation (Adams 1958; Denker and Gerdes 1979; Leiser and Denker 1988; Fischer et al. 1991).

Fertilization occurs shortly after ovulation in the ampullar region of the oviducts by sperm that have already been present for 10–12 h and capacitated in the female genital tract (Chang 1951). The resulting zygote progresses through the next stage of development rather quickly. Pronuclear formation takes about 6 h. The cell-doubling time is about 6 h for the first two divisions, all occurring under maternal control. Overlap of cell stages among different embryos is commonly observed, especially following superovulation (Varian et al. 1967).

The switch from maternal to zygotic control of development occurring between the 8–16 cell stages is similar to this in cattle and sheep (Barnes and Eyestone 1990; Telford et al. 1990). It is characterized by the loss or decay of mRNA of maternal origin, activation of transcription of the embryonic genome, developmental arrest of the embryo in the presence of transcription inhibitors, also by the marked qualitative changes in protein synthetic patterns in the embryos (Telford et al. 1990). Cell-doubling time is about 8 h after the transition until compaction is initiated at 32–64 cell stages (Ziomek et al. 1990).

Embryos reach the uterus between 60 and 72 h after ovulation. However, in superovulated rabbits this process can be delayed (Yang 1991). Blastocoele formation starts around the 128-cell stage with an increased cell doubling time of 12 h (Stice and Robl 1988; Ziomek et al. 1990). By this time, the mucin layer and the zona pellucida are less resistant to enzymatic removal (Ziomek et al. 1990) and to mechanical penetration by using a micro-pipette (Yang and Foote 1987). Blastocyst expansion and zona pellucida dissolution start around 84 h post-ovulation. In *in vivo* environment, the blastocysts do not undergo a real hatching process due to the presence of the mucin coat. The disappearance of zona pellucida and mucin coat occurs approximately 96 and 140 h post ovulation, respectively (Adams 1958; Fischer et al. 1991).

The secreted proteins by blastocysts form a new layer of coating prior to implantation under the mucin coat. This new layer is called neo-zona. Its formation does not take place under the non-physiological *in vitro* developmental culture conditions (Fischer et al. 1991). Implantation of rabbit embryos takes place around 7 days following artificial insemination (Orsini 1962).

The average gestation lengths are 30–31 days. They vary slightly depending on the breed, and on litter size. Minimal number of corpora lutea is required for successful pregnancy maintenance and it varies depending on the breed (Beatty 1958, Feussner et al. 1992). Most of the post-implantation losses occur during the two critical periods in rabbit pregnancy. Adams (1960) observed that among the approximate 18% post-implantation losses, 7% happened immediately post-implantation, 66% between days 8 and 17, and 27% between days 17 and 23. The first critical period, around day 13, is related to the placentation changes from the yolk sac to the hemochorial type. The second critical period falls between days 22–23. The tense, rounded fetal structures are susceptible to dislodgment during this time interval. The placenta becomes hemoendothelial around day 23, separating the maternal and fetal circulation with only one layer.

10.3 State of Nuclear Transfer

The procedure of rabbit nuclear transfer is similar to this used in other species regarding to oocyte and somatic cell donors (Fig. 10.1). Detailed discussion explains the important steps of the procedure emphasizing the special aspects unique to rabbits in the following section.

10.3.1 The Recipient Oocytes and Their Enucleation

In other species ***in vitro* matured** oocytes are commonly used for cloning (Lonergan et al. 2000; Kasinathan et al. 2001). However, it is rare in the case of rabbit (Yin et al. 2002a). Most rabbit nuclear transfer experiments are using *in vivo* matured and ovulated oocytes. They are usually collected by flushing the oviducts (Maurer 1978). Variety of **superovulation** protocols can be used to increase oocyte production, and approximately 30–40 eggs per doe can be produced (Kennelly and Foote 1965; Maurer et al. 1968; Carney and Foote 1990; Yang et al. 1990a, b; Kauffman et al. 1998). The response to the superovulation treatment with regard to the number and quality of the oocytes is varied among different breeds. Superovulation protocols can also affect the quality of the resulting embryos, including their cell numbers, their sensitivity to cryopreservation and the subsequent pregnancy rates (Carney and Foote 1990; Kauffman et al. 1998). The **age of the oocytes** has also an important role in the subsequent progress. The highest efficiency was achieved while the

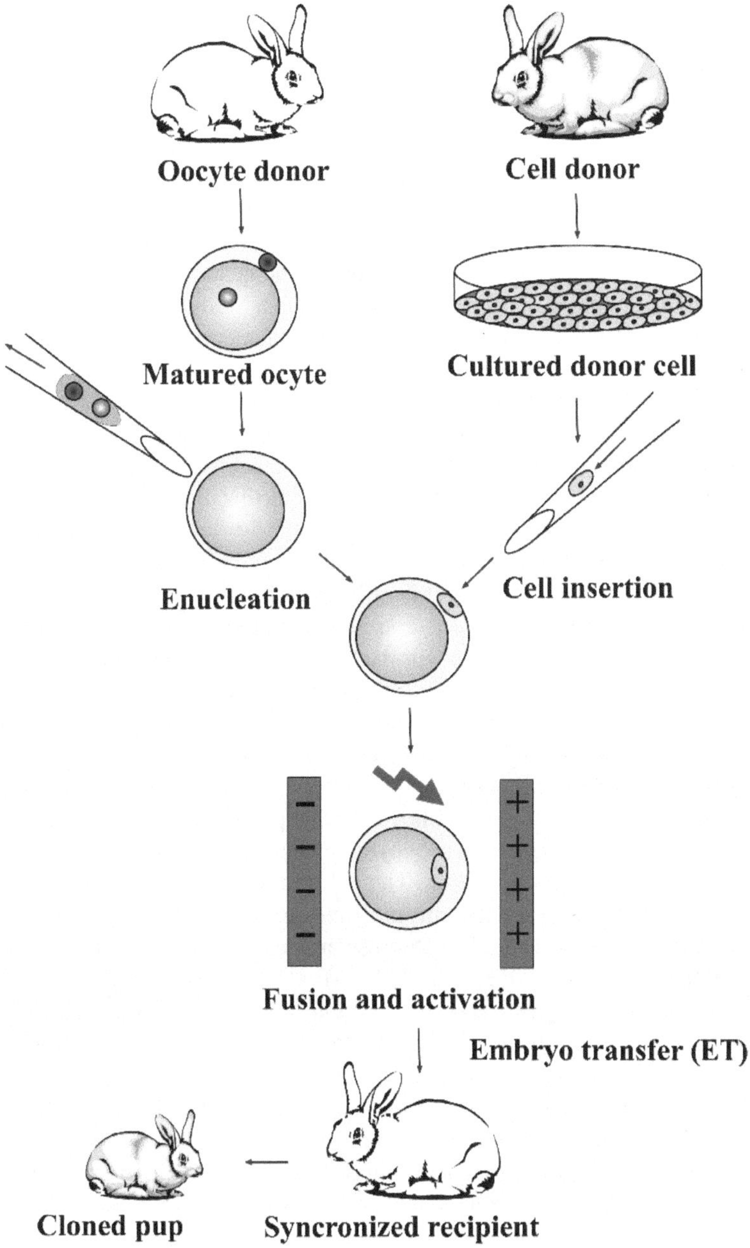

Fig. 10.1 Procedure of the rabbit somatic cell nuclear transfer

16h post hCG oocytes were collected and manipulated (Collas and Robl 1990). Recently ovulated oocytes have been found to have higher fusion and development rate than the older oocytes, although the older ones activate better (Cervera and Garcia-Ximénez 2003). Furthermore, the conditions of the aging are also important for the capacity of oocyte to support development of the donor nucleus, besides the post ovulatory age. The *in vitro* aged oocyte cytoplasm approaches an interphasic stage and reduces asynchrony between the donor cell and the recipient environment (Adenot et al. 1997).

The **breed of the oocyte donors** may also have a major effect on the outcome of nuclear transfer, as previously demonstrated in mouse (Wakayama and Yanagimachi 2001) and in sheep (Dinnyes et al. 2001b). Only a few rabbit breeds were tested in nuclear transfer experiments (Mitalipov et al. 1999; Chesné et al. 2001; Dinnyes et al. 2001a). Furthermore, interactions between the genetic origin of the recipient oocytes and the donor cells are yet to be investigated. Studying these variables may shed lights on the contributing factors to the lack of success in the development of cloned embryos in several studies.

Cryopreserved oocytes represent another resource to produce cytoplasts for nuclear transfer as described for cattle (Dinnyes et al. 2000). Rabbit oocytes are not particularly sensitive to low temperature. Cryopreserved oocytes were successfully used to produce progenies following *in vitro* fertilization (Vincent et al. 1989). However, cryopreserved rabbit oocytes had not been tested in nuclear transfer experiments.

The efficiency of nuclear transfer is influenced by other circumstances besides the oocyte. The **enucleation** is a complicated step of this technology. The enucleation of oocytes in rabbit is similar to those of other species. The original procedure is to remove the nucleus with a sharp glass pipette. The zona pellucida and the cell membrane are neither too hard nor too flexible to cause difficult conditions during micromanipulations. Depending on the breed of oocyte donors, Nomarski optics usually allows the visualization of the metaphase chromosomes in most of the oocytes, and their removal can be performed under visual control. Morphology and quality of rabbit oocytes, due to variability, often reduce the efficiency of the micromanipulation. The coloration and granulation of the cytoplasm can prevent one to see the chromosomes using standard light microscopy, and may necessitate the use of epifluorescent staining and UV exposure of the oocytes (Dinnyes et al. 2001a). Similarly, Mitalipov et al. (1999) also reported that Dutch Belted's oocytes were more difficult to manipulate than those of New Zealand Whites', due to the darkness and opacity. In some oocytes, the metaphase plate and the polar body are often not in proximity, therefore, in many cases blind enucleation is not efficient to remove the nuclear material, and epifluorescent staining was necessary to complete the enucleation. Furthermore, fragmentation of oocytes of some donors can reach as high as 100% shortly after micromanipulation, but others from the same treatment groups showed no fragmentation or just low levels (Dinnyes and Yang, 1997). The enucleation conditions, especially the composition of the media, have been shown to have a major effect on further development as described by Collas and Robl (1990).

The enucleation can be completed using a chemically-assisted method. In this case ionomycin and demecolcine induce membrane protrusion that contains all the maternal chromosomes which could be easily removed. The time of the enucleation usually occurs before donor cell insertion and fusion. However, a few years ago delayed enucleation was described when the enucleation occurred 1 or 2 h after fusion. Although, this method was more efficient and it caused slightly higher implantation rate, following the publication it was not commonly applied (Yin et al. 2002b).

10.3.2 The Donor Cell

The donor cell type (embryonic or adult somatic cell), cell cycle (G0/G1 or S), culture condition (fresh or cultured) and the introduction to the recipient oocyte (perivitellinar space or into the cytoplasm) are also very important parts of the nuclear transfer procedure.

10.3.2.1 Origin (cell type)

In the beginning of the rabbit cloning history, early stage embryo blastomers were used as donor cells (eight cell blastomers: Stice and Robl 1988, 8-16-32 cell blastomers: Yang et al., 1990, 8-16 cell blastomers: Collas and Robl 1990, morula: Yang et al. 1992). The efficiency of producing cloned live offspring was from 3% to 10%. Adult rabbit fibroblast (Mitalipov et al. 1999) and cumulus cells (Yin et al. 2000) were used from the end of 1990s, but no live pups were obtained till 2002 (Fig. 10.2). Live cloned rabbit offspring were produced from somatic cells using cumulus cells in 2002 (Chesné et al. 2002) and then from adult fibroblasts (Li et al. 2006). These results are in agreement with observations on *in vitro* development of blastocyst with nuclear donor cells such as cumulus cells (27%) vs. fibroblasts (3%) (Cervera and Garcia-Ximénez 2003). Development to term of SCNT rabbits from cumulus cells confirms the difference between cell types in recent reports (Meng et al. 2008).

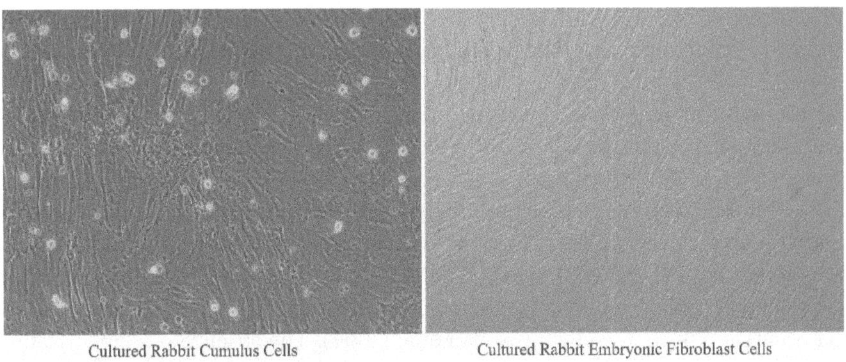

Fig. 10.2 Cultured somatic cells of rabbit

10.3.2.2 Cell Cycle

The influence of the donor cell cycle on nuclear transfer embryos is also well studied in various species, including rabbits (Stice et al. 1993). Collas and colleagues (1992a) studied the effect of the donor cell cycle stage on the nuclear transferred embryo development. Reconstructed embryos from G1 (59%) and early S (32%) stage donor cells could develop better to blastocyst stage than embryos from late S (3%). These results were explained with the abnormalities of the metaphase chromosomes, incomplete or absent spindle formation and incomplete chromatin condensation when late S stage donor cells were used (Collas et al. 1992b).

The effects of somatic cell cycle synchronization on the success of nuclear transfer were described with special emphasis on the usage of G0-stage cells, which was found to be beneficial in the first successful somatic cell cloning experiments (Wilmut et al. 1997). The cell cycle could be affected by chemicals (nocodazol, Piotrowska et al. 2000) and also by culture conditions: treatments of serum starvation or contact inhibition (confluency) arresting the majority of cells in either G0 or G1 stage, respectively (Boquest et al. 1999; Dinnyes et al. 2001a). However, the accurate evaluation of the cell-cycle effect on the development of individual embryos is difficult, because a given donor cell cannot be analyzed. Due to the low efficiency of nuclear transfer at the current state, rare events may possibly lead to the production of the 2–5% cloned progeny. Therefore statistical probabilities may not provide the best estimates of the analysis of donor cells for somatic cell nuclear transfer.

In rabbits, cloned blastocysts have been obtained from serum-starved G0-stage adult fibroblasts (Mitalipov et al. 1999), and serum-starved G0/G1-stage cultured cumulus cells (Yin et al. 2000). *In vitro* development to the blastocysts stage was not different in embryos cloned from fibroblast cells serum-starved vs. contact inhibited (Dinnyes et al. 2001a; Li et al. 2006). Galat et al. (1999) compared quiescent and cycling fetal fibroblast cells as nuclear donors and observed similar cleavage development. However, increased blastocysts rates were obtained using cells subjected to serum-starvation treatment. Cloned progeny were obtained with fresh cumulus cells, naturally in a resting stage (Chesné et al. 2002; Meng et al. 2008).

10.3.2.3 Epigenetic Status (histone acetylation, methylation)

The **methylatio**n and histone acetylation status of the cells are in focus of recent investigations, besides the donor cell cycle. During early mammalian development, dramatic chromatin structure changes and nuclear reorganization are occurring. During preimplantation development in most of the mammalian species embryos undergo early stage active demethylation and remethylation events. In contrast with mouse, rat, pig and cattle, the rabbit embryo exhibits equally high methylation levels genome-wide from the zygote to the 16-cell stage and the passive demethylation occurs in 16- to 32-cell stages (Shi et al. 2004). The absence of demethylation mechanism could explain the low number of successes in rabbit NT. Additionally, aberrant methylation patterns were observed in NT rabbit embryos, which could disturb zygotic gene activation and might lead to subsequent developmental failures and embryo death (Chen et al. 2004).

The epigenetic status of the donor cell could be described by the level of **histone acetylation**. Recently, the histone acetylation levels of cumulus and fibroblast donor cells, and their effects on the cloning procedure (Yang et al. 2007) were evaluated. The study revealed that the cultured cumulus cells have had higher acetylation level than that of the fetal fibroblast cells. Although this had no influence on the *in vitro* development rate, it might have contributed to the higher implantation rate and full term development. The donor cells were treated with sodium butyrate (NaBu), a histone deacetylase (HDAC) inhibitor, to increase the acetylation level. Although the treatment was successful and increased the blastocyst rates, still lower pregnancy rates and no birth of offspring were detected.

Similarly, the treatment with TSA, another HDAC inhibitor, improved the cloning efficiency in cattle (Enright et al. 2003) and in mice (Kishigami et al. 2006; Rybouchkin et al. 2006; Wang et al. 2007). Moreover, recent reports indicated that TSA treatment could increase the cell number of rabbit SCNT blastocysts (Xu et al. 2007) or the blastocyst development rate (Shi et al. 2008a) and the histone acetylation pattern of TSA-treated rabbit SCNT embryos appeared to be more similar to that of normal embryos (Shi et al. 2008b). Recently, term development of TSA-treated cloned embryos has been achieved by our group (Meng et al. 2008), with no difference in the pregnancy or birth rate between TSA-treated and untreated SCNT embryos as both could develop to term. The postnatal growth rates of these two groups did not differ, but only the TSA untreated pups survived to adulthood. These surviving SCNT rabbits were bred further successfully.

10.3.3 Fusion of Donor Cells and Activation

The activation is one of the most important steps of SCNT. In normal fertilization the sperm penetration induces repeated rises of calcium ion levels in the cytoplasm, which results in the decrease of the level of maturation promoting factor (MPF). In NT, this sperm-generated effect is replaced by artificial activation. The usual triggers of activation are electric pulse(s) and/or a short exposure to chemical agents. Collas and Robl (1990) showed the advantages of multiple electric pulses instead of a single stimulus. To mimic the calcium oscillation, high concentration of inositol 1, 4, 5-triphosphate (IP3) was reported in several publications leading to a better development rate and a post-implantation development (Mitalipov et al. 1999; Inoue et al. 2002). After fusion and electrical activation, the exposure to cytochalasin B improved the efficiency of NT. This inhibitor helps to decrease the deficiency of microfilament-dependent cytoskeletal mechanism under activation (Yang et al. 1992; Li et al. 2002). 6-dimethylaminopurine (DMAP), a kinase phosphorylation inhibitor, was applied in rabbit cloning successfully inducing pronuclear formation (Mitalipov et al. 1999; Dinnyés et al. 2001a). To improve developmental capacities, various electrical pulses and chemical treatments were combined. The addition of cycloheximide (CHX), a protein synthesis inhibitor, could increase the NT embryo development resulting live pups (Chesné et al. 2002; Liu et al. 2004).

10.3.4 Embryo Culture and Cryopreservation

Significant progress has been made during recent years in *in vitro* culture of rabbit zygotes (Kane and Foote 1971; Maurer 1978). Using either co-culture (Carney and Foote 1990; Carney et al. 1990), media containing peritoneal fluid (Collas et al. 1991), vitreous humor of the eyes of female rabbits (Collas and Robl 1991) or even completely defined protein-free systems (Carney and Foote 1991; Li et al. 1993; Farrell and Foote 1995; Li and Foote 1996; Giles and Foote 1997), high rates of blastocyst development have been obtained. In the case of SCNT rabbits, the embryos can develop in a number of different culture media (M199, PBS, EBSS, VH, RD, B2) at 37–39°C, under 5% CO2 in air with 95% humidity following reconstruction and activation. Menezo B2 medium was successfully used to culture parthenogenetically activated oocytes to the blastocyst stage (Ozil 1990), and SCNT embryos prior to transfer into recipients (Chesné et al. 2002). Both, TCM 199 and KSOM media were successfully used to culture rabbit embryos from SCNT (Dinnyés et al. 2001a; Chesné et al. 2001). Although both media support high rates of *in vitro* blastocyst development, a short culture in KSOM resulted in superior *in vivo* embryo survivals (B. Wang and X. Yang 1994). Our experience shows that only a short period of *in vitro* culture is advisable to increase the chance of obtaining progeny prior to embryo transfer. Mitalipov et al. (1999) found that Earle's balanced salt solution (EBSS) supplemented with MEM nonessential amino acids, basal medium Eagle amino acids, L-glutamine, sodium pyruvate and fetal bovine serum (FBS) was able to support blastocyst development of activated oocytes better than with Dulbecco's modified Eagle's medium (DMEM/RPMI), FBS, CR1aa or FBS. They also obtained a good blastocyst development rate (30%) with embryos from somatic cell nuclear transfer using the same medium. Similar results were also obtained with the EBSS system by Yin et al. (2000) (Fig. 10.3).

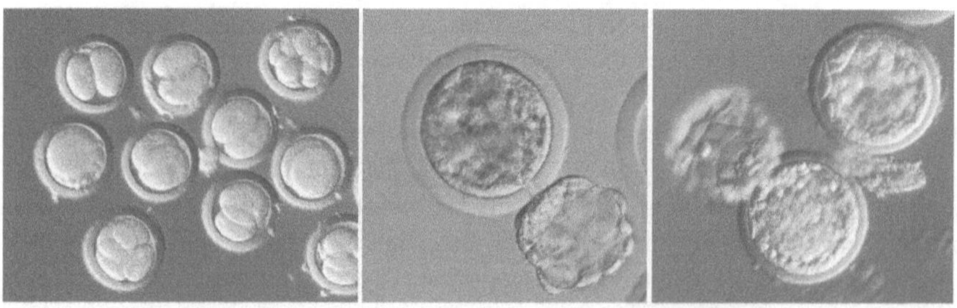

Early stage of SCNT Rabbit embyos SCNT Rabbit blastocysts

Fig. 10.3 *In vitro* cultured SCNT rabbit preimplantation stage embryos in EBSS

Embryo cryopreservation, another important technique for practical embryology research, is fairly successful in rabbits (Tsunoda et al. 1982; Renard et al. 1984; al-Hasani et al. 1989; Vincent et al. 1989; Kasai et al. 1992). Blastomeres from frozen rabbit morulae were successfully used as nuclear donors, resulting in the birth of young (Heyman et al. 1990). Furthermore, cloned embryos from blastomer nuclear transfer were cryopreserved/thawed before being used as donors in a second round of nuclear transfers (Rao et al. 1998).

10.3.5 Embryo Transfer, Pregnancy Monitoring and Progeny Production

Rabbit embryos accumulate a mucin coat during their passage through the oviduct, and without this mucin layer, their viability is limited (Murakami and Imai 1996). Although successful progeny development was recently reported (Jin et al. 2000) following prolonged *in vitro* culture and blastocyst-stage embryo transfer, rabbit embryos need to be transferred shortly after micromanipulation to obtain high number of progeny. Embryos are usually surgically transferred into the oviduct, by midventral laparotomy (Yang and Foote 1990). Laparoscopic transfer to the Fallopian tubes also offers an efficient and faster alternative for early and late cleavage stage embryos (Besenfelder and Brem 1993; Besenfelder et al. 1998). Recently, successful non-surgical transfer of later stage embryos through the cervix into the uterus was also reported (Kidder et al. 1999).

The synchrony between the recipient oviduct or uterus and the transferred embryos is an important issue in rabbits (Chang 1951) and it was studied extensively with *in vitro* produced embryos. Asynchronous transfer can result in various abnormalities and death of the embryos. Rabbit blastocysts are more sensitive to asynchrony than earlier cleavage stage embryos (Fischer 1989). Depending on the age of the embryo, synchronous or +/– 1 day asynchronous transfers gave the highest fetal development (Tsunoda et al. 1982; Fischer 1989; Yang and Foote 1990). A minimum of two implanted embryos seems to be necessary to carry a pregnancy to term (Adams 1970). Conversely, overcrowding of the uterus by transfer of too many embryos can reduce embryonic and fetal survival (Hafez 1964). This was demonstrated in a report by Besenfelder and Brem (1993) who obtained a higher rate of progeny survival following transfer of 10–20 embryos as compared to 30–65 embryos. Most interestingly, synchronous embryo transfer with cloned embryos failed to establish pregnancy. However, a relatively large asynchrony (22-h) was one of the major factors for the success in the first production of somatic cell cloned rabbits (Chesné et al. 2002).

Rabbit pregnancy can be detected by palpation around day 10–14 after artificial insemination (Harkness and Wagner 1995). During certain periods of pregnancy (around day 23, as discussed earlier), the fetuses are very sensitive to handling, and palpation itself can cause abortion.

Embryos from blastomere nuclear transfer have resulted pregnancies and healthy progeny (Stice and Robl 1988; Collas and Robl 1990; Heyman et al. 1990;

Newborn SCNT Rabbit 1week old SCNT Rabbit 1,5 years old SCNT Rabbit

Fig. 10.4 Post-natal development of the first Hungarian SCNT rabbit (Tapsilla)

Yang et al. 1992). Full term development, however, is low due to the high incidence of post-implantation losses (Heyman et al. 1990; Heyman and Renard 1996). Birth of abnormal young has also been reported (Park et al. 1998).

Somatic cell nuclear transfer can often result in high pregnancy losses and a variety of fetal and perinatal abnormalities as shown in cattle and in sheep (Dinnyés et al. 2008). The frequency of the abnormalities depends on many factors. Species differences are likely among them, as reports on progeny from goat and pig somatic cell nuclear transfer showed that the majority of them are healthy during late pregnancy and after birth (Polejaeva et al. 2000; Betthauser et al. 2000; Baguisi et al. 1999). The first litters of somatic cell cloned rabbits were normal in their morphological appearance and the loss of two of six young can be contributed to the failure in the adoptive process from lactating mother. The four others developed normally and two of them proved to be fertile (Chesné et al. 2002).

Our recent results demonstrated that among cumulus cell derived clones, only 4 out of the 13 live born progeny reached adulthood and all of them were fertile (Meng et al. 2008) (Fig. 10.4). However, fibroblast cell derived live born have not been reported to reach fertile adulthood (Li et al. 2006).

10.4 Possible Applications of Rabbit Cloning

The somatic cell nuclear transfer (SCNT) is a powerful tool for research, although, the efficiency of SCNT animal production is still low in most species. Considerable possibilities of cloning exist for research, biomedical and agriculture applications (Vajta and Gjerris 2006).

Most of the current interest is in the biomedical field. The SCNT combined with transgenic and/or embryonic stem (ES) cells technology could offer new biomedical animal models for some human diseases. Since the 1980s it has been possible to genetically modify mammals by microinjection of designed DNA into the pronuclear stage zygotes. Unfortunately, this method is of limited efficiency because of the poor development of injected embryos, the random transgene integration and the variable numbers of the transgene copies. Transgenic animal can be created with genetically modified ES cells. However germ line transition only occurred in mouse. Several transgenic animals are designed to mimic human diseases which can be used to further understand the diseases and to experiment possible treatments. Numerous human disease models have been developed in mice. Unfortunately not all of the human diseases can be modeled with mice due to the limitations of the physiological differences between mouse and human. SCNT is an efficient way to produce genetically modified animals and it may thus avoid these problems. Specific gene targeted somatic cells (ES or other somatic cells) can be introduce into enucleated oocyte by nuclear transfer and produce viable transgenic animals. The cloning efficiency is also low but the created transgenic animals may have high added value. Non-rodent laboratory (for example rabbits) or farm animals (for example sheep and pigs) can be more appropriate models due to some physiological similarities to human, size of their organs and lifetime. The advantage of rabbit is its dual status of small laboratory and farm animal, being large enough to evaluate a number of diseases. Its vascular system is similar to human. Moreover, it is easy to keep rabbits in larger numbers in well defined laboratory conditions, to breed them as they have a relatively short reproduction cycle. The rabbit can also be used as a bioreactor. Bioreactors are transgenic animals which have human genes integrated into their genome to produce humanized proteins. Usually, these proteins are expressed in the milk. Although there are several farm animals producing milk, rabbits have a particularly high level of proteins in their milk. Transgenic rabbits have been extensively used for production of therapeutic proteins in their mammary gland. Milking of the does is relatively simple (Lebas 1970; Marcus et al. 1990). Milk production varies significantly according to breeds, and as much as 200 ml per day can be obtained in New Zealand Whites.

In agriculture there are two options to use cloning in farm animals. One is to use reproductive cloning to create copies of animals with high value traits, like a dairy cow with high milk production or bulls with high quality meat. The other option consists of introducing genetic modifications for beneficial traits, including disease resistance, adaptation to harsher environments, reduced waste production or products with novel valuable phenotypes. Several resistances (Wall et al. 2005) and valuable genes such as specific casein genes in cattle (Brophy et al. 2003) could improve the value of the animals. However it is not clear yet whether in rabbits which have a low individual value compared to bovine and to pig, this technology would be viable economically.

Finally, an important use of SCNT is for basic research. Cloning of farm animals could help to understand the basic biological mechanisms, including gene expression involved in early embryo development and differentiation (Santos and Dean 2004). Numerous interesting questions are still open in the field of epigenetics and the nuclear transfer is one of the most suitable techniques to answer them. Rabbits could also be the subject of this research and several studies were already published focusing on the process of nuclear reprogramming (Pinto-Correia et al. 1995; Shi et al. 2004).

10.5 Appendix: Protocol for Rabbit Somatic-Cell Nuclear Transfer

The protocol of rabbit somatic-cell nuclear transfer is provided below. Many other alternatives exist for almost every step, including those which are more efficient. In the light of recent successes, the activation procedure and other supplemented treatments including TSA treatment, embryo transfer, co-transfer with PGA embryos and the synchronization steps need special attention in this species. This protocol resulted in 13 live pups (four reached the adulthood and produced healthy progeny) from cumulus cells, and it resulted also in 46.6% development of blastocysts from fibroblast. However there were no offspring. All chemicals used were purchased from Sigma Chemical Co., Invitrogen and Gibco.

10.5.1 Superovulation of Oocyte Donors and Oocyte Collection

Mature (20–22 weeks old) Hycole hybrid female rabbits were superovulated by injection of 120 IU pregnant mare serum gonadotrophin (PMSG, Folligon, International B.V., Boxmeer, Holland) intramuscularly and 180 IU human chorionic gonadotrophin (hCG Choragon, Ferring GmbH, Kiel, Germany) intravenously 72 h later. Mature oocytes were flushed from the oviducts 13–14 h post-hCG injection with Medium 199 supplemented with 10% Fetal Calf Serum (FCS, Gibco 10108-165) and 20 mM Hepes (Sigma H4034). Cumulus cells were removed from oocytes by gentle pipetting into M199 containing 5 mg ml^{-1} hyaluronidase (Sigma H3506) at 37°C, and incubated in Earles Balanced Salt Solution (Sigma E2888)-complete (EBSS-complete) (Mitalipov et al. 1999) with 5% CO_2 in air at 38.5°C until use.

10.5.2 Enucleation of the Oocytes

The somatic cell nuclear transfer procedure is carried out essentially as described previously (Challah-Jacques et al. 2003) with minor modifications. Briefly, the denuded oocytes were enucleated by micromanipulation. The oocytes were stained with 5 μg ml^{-1} Hoechst 33342 in EBSS-complete media for 20 min. After pretreatment with 7.5 μg ml^{-1} cytochalasin B (Sigma C6762) in M199 for 10 min, the metaphase plate and the first polar body were located applying UV illumination for 1–2 s and were removed by a 18–20 μm outer diameter micropipette. The enucleated oocytes were allowed to recover in EBSS-complete media for 2 h prior to cell insertion.

10.5.3 Donor Cell Preparation

When the cumulus cells were used as nuclear donors, they were collected as above from the cumulus-oocyte-complexes of one donor female, and centrifuged at 3,000 rpm for 1 min and then kept in M199 supplemented with FCS and Hepes at 4°C prior to cell insertion.

When fibroblast cells were used as nuclear donors, they were collected from an ear skin biopsy of an adult transgenic rabbit. The biopsy area was shaved and the surface was cleaned with 70% ethanol. A small piece of tissue was cut from the ear and washed several times in PBS with antibiotic solution (Pen/Strep, Invitrogen 15140-122), then cut into 1 mm cubes, and placed into a Petri dish to culture as explants with DMEM (Invitrogen, 31966-021) medium with 10% FCS. After approximately 14 days, the fibroblast cells, growing out of these explants, were washed twice with Ca^{2+} and Mg^{2+} free PBS then harvested by 0,05% trypsin EDTA (Invitrogen, 25300-054). These cells were washed by centrifugation, resuspended, then used for nuclear transfer (referred to as non-passaged or passaged cells) or cultured further for up to 15 passages in DMEM with 10% FCS. Nuclear donor cells were isolated from non serum-starved culture drops supporting a fully confluent cell monolayer for 2–5 days prior to the experiment. Alternatively, serum starved cells were obtained by exposing fully confluent cell cultures to 0.5% FCS in DMEM for 3–5 days. Cell monolayers were trypsinized, and cells were washed by centrifugation in DMEM + 10% or 0.5% FCS, respectively, then incubated at 38.5°C in drops and used within 1 h.

10.5.4 Nuclear Transfer

A cumulus or fibroblast cell was inserted in the perivitelline space of the enucleated oocyte with the micropipette. The cytoplast-cell construct was induced to fuse by three 20 µs 3.2 kVcm^{-1} DC pulses in activation medium (0.25 M sorbitol (Sigma, S3889) in water supplemented with 0.5 mM Hepes, 0.1 mM $Ca(CH_3COO)_2$, 0.5 mM $Mg(CH_3COO)_2$ and 1 mg ml^{-1} bovine serum albumin (Sigma, A3311). Fused embryos were activated by the same parameter electrical pulses 1 h later and treated with 2 mM 6-dimethylaminopurine (6-DMAP), (Sigma, D2629) and 5 µg ml^{-1} cycloheximide (CHX) (Sigma, C1988) in EBSS-complete media for 1 h. The cloned embryos were then subsequently cultured in EBSS-complete media with or without 5 nM Trichostatin A (TSA) (Sigma, T8552) for 10 h as described before (Kishigami et al. 2006; Xu et al. 2007).

10.5.5 Embryo Culture

The nuclear transfer embryos were cultured for 4.5 days in 20 µl drops of EBSS-complete media under oil in a humidified atmosphere of 5% CO_2 in air at 38.5°C either overnight until 2- to 4-cell or for 4.5 days until expanded blastocyst stage. Blastocyst development was recorded and blastocyst cell numbers were counted following Hoechst 33,342 fluorescein staining.

10.5.6 Embryo Transfer

Embryos were transferred at 2- to 4-cell stage after overnight culture in EBSS complete media as described above. The 10–15 embryos were transferred through the infundibulum into each oviduct of recipients using a laparoscopic technique as described before (Besenfelder and Brem 1993). All recipient does were administered with 0.2 ml GnRH analogue (Receptal®, Intervet) at 22 h after injecting the oocyte donor females (22 h asynchrony with SCNT embryos) to induce ovulation. Recipients were operated by caesarean sections on day 30 post-ET. The birth and placental weights of progeny were measured, and the live pups were fostered to a female with similar age pups. Some recipients were allowed to give birth naturally 32 days post-embryo transfer; the live pups were then raised by their recipient mothers.

Acknowledgements This manuscript was supported by Wellcome Trust (Grant No.070246), Teamoholic (MEXT-CT-2003–509582), CLONET (MRTN-CT-2006–035468), MED-RAT (LSHG-CT-2006–518240), and Chinese-Hungarian Bilateral projects (TET CHN-28/04, CHN-56/07).

Bibliography

Adams, C. E. (1958) Egg development in the rabbit. The influence of post-coital ligation on the uterine tube and ovariectomy. J. Endocrinol. 16: 283–294.

Adams, C. E. (1960) Prenatal mortality in the rabbit Oryctolagus cuniculus. J. Reprod. Fertil. 1: 36–44.

Adams, C. E. (1970) Maintenance of pregnancy relative to the presence of few embryos in the rabbit. J. Endocrinol. 48: 243–249.

Adams, C. E., Hay, M. F., and Lutwak-Mann, C. (1961) The action of various agents upon the rabbit embryo. J. Embryol. Exp. Morphol. 9: 468–491.

Adenot, P. G., Szöllösi, M. S., Chesné, P., Chastant, S., and Renard J. P. (1997) In vivo aging of oocytes influences the behavior of nuclei transferred to enucleated rabbit oocytes. Mol. Reprod. Dev. 46(3): 325–336.

al-Hasani, S., Kirsch, J., Diedrich, K., Blanke, S., van, d., V, and Krebs, D. (1989) Successful embryo transfer of cryopreserved and in-vitro fertilized rabbit oocytes. Hum. Reprod. 4: 77–79.

Anderson, J. and Henck, J. W. (1994) Toxicity and Safety Testing. In"The Biology of the Laboratory Rabbit" (Manning, P. J., Ringler, D. H., and Newcomer, C. E., Eds.), pp. 455–466, Academic, San Diego, CA.

Baguisi, A., Behboodi, E., Melican, D. T., Pollock, J. S., Destrempes, M. M., Cammuso, C., Williams, J. L., Nims, S. D., Porter, C. A., Midura, P., Palacios, M. J., Ayres, S. L., Denniston, R. S., Hayes, M. L., Ziomek, C. A., Meade, H. M., Godke, R. A., Gavin, W. G., Overstrom, E. W., and Echelard, Y. (1999) Production of goats by somatic cell nuclear transfer. Nat. Biotechnol. 17: 456–461.

Barnes, F. L. and Eyestone, W. H. (1990) Early cleavage and the maternal zygotic transition in bovine embryos. Theriogenology 33: 141–152.

Barnes, F. L., Collas, P., Powell, R., King, W. A., Westhusin, M., and Shepherd, D. (1993) Influence of recipient oocyte cell cycle stage on DNA synthesis, nuclear envelope breakdown, chromosome constitution, and development in nuclear transplant bovine embryos. Mol. Reprod. Dev. 36: 33–41.

Beatty, R. A. (1958) Variation in the number of crpora lutea and in the number and size of 6-day blastocysts in rabbits subjected to superovulation treatment. J. Endocrinol. 17: 248–260.

Besenfelder, U. and Brem, G. (1993) Laparoscopic embryo transfer in rabbits. J. Reprod. Fertil. 99: 53–56.

Besenfelder, U., Strouhal, C., and Brem, G. (1998) A method for endoscopic embryo collection and transfer in the rabbit. Zentralbl. Veterinarmed. A 45: 577–579.

Betthauser, J., Forsberg, E., Augenstein, M., Childs, L., Eilertsen, K., Enos, J., Forsythe, T., Golueke, P., Jurgella, G., Koppang, R., Lesmeister, T., Mallon, K., Mell, G., Misica, P., Pace, M., Pfister-Genskow, M., Strelchenko, N., Voelker, G., Watt, S., Thompson, S., and Bishop, M. (2000) Production of cloned pigs from in vitro systems. Nat. Biotechnol. 18: 1055–1059.

Boquest, A. C., Day, B. N., and Prather, R. S. (1999) Flow cytometric cell cycle analysis of cultured porcine fetal fibroblast cells. Bio. Reprod. 60: 1013–1019.

Brophy, B., Smolenski, G., Wheeler, T., Wells, D., L'Huillier, P., Laible, G. (2003) Cloned transgenic cattle produce milk with higher levels of β-casein and κ -casein. Nat. Biotechnol. 21: 157–162.

Brunet-Simon, A., Henrion, G., Renard, J. P., and Duranthon, V. (2001) Onset of zygotic transcription and maternal transcript legacy in the rabbit embryo. Mol. Reprod. Dev. 58: 127–136.

Campbell, K. H., Loi, P., Otaegui, P. J., and Wilmut, I. (1996) Cell cycle co-ordination in embryo cloning by nuclear transfer. Rev. Reprod. 1: 40–46.

Campbell, K. H. S., McWhir, J., Ritchie, W. A., and Wilmut, I. (1996) Sheep cloned by nuclear transfer from a cultured cell line. Nature 380: 64–66.

Capecchi, M. (2000) How close are we to implementing gene targeting in animals other than the mouse. Proc. Natl. Acad. Sci. USA 97: 956–957.

Carney, E. W. and Foote, R. H. (1990) Effects of superovulation, embryo recovery, culture system and embryo transfer on development of rabbit embryos in vivo and in vitro. J. Reprod. Fertil. 89: 543–551.

Carney, E. W. and Foote, R. H. (1991) Improved development of rabbit one-cell embryos to the hatching blastocyst stage by culture in a defined, protein-free culture medium. J. Reprod. Fertil. 91: 113–123.

Carney, E. W., Tobback, C., Ellington, J. E., and Foote, R. H. (1990) Co-culture of rabbit 2-cell embryos with rabbit oviduct epithelial cells and other somatic cells. Mol. Reprod. Dev. 27: 209–215.

Cervera, R. P. and García-Ximénez, F. (2003) Oocyte age and nuclear donor cell type affect the technical efficiency of somatic cloning in rabbits. Zygote 11(2): 151–158

Challah-Jacques, M., Chesne, P., and Renard, J. P. (2003). Production of Cloned Rabbits by Somatic Nuclear Transfer. Cloning Stem Cells. 5: 295–299

Chang, M. C. (1951) Fertilization capacity of sperm deposited in the fallopian tube. Nat. (Lond.) 168: 697.

Chang, M. C., Casas, J. H., and Hunt, D. M. (1971) Development of ferret eggs after 2 to 3 days in the rabbit fallopian tube. J. Reprod. Fertil. 25: 129–131.

Chen, T., Zhang, Y. L., Jiang, Y., Liu, S. Z., Schatten, H., Chen, D. Y., and Sun, Q. Y. (2004) The DNA methylation events in normal and cloned rabbit embryos. FEBS Lett. 578: 69–72.

Chen, T., Zhang, Y. L., Jiang, Y., Liu, J. H., Schatten, H., Chen, D. Y., and Sun, Q. Y. (2006) Interspecies nuclear transfer reveals that demethylation of specific repetitive sequences is determined by recipient ooplasm but not by donor intrinsic property in cloned embryos. Mol. Reprod. Dev. 73(3): 313–317.

Chen, Y., He, Z. X., Liu, A., Wang, K., Mao, W. W., Chu, J. X., Lu, Y., Fang, Z. F., Shi, Y. T., Yang, Q. Z., Chen da, Y., Wang, M. K., Li, J. S., Huang, S. L., Kong, X. Y., Shi, Y. Z.,

Wang, Z. Q., Xia, J. H., Long, Z. G., Xue, Z. G., Ding, W. X., and Sheng, H. Z. (2003) Embryonic stem cells generated by nuclear transfer of human somatic nuclei into rabbit oocytes. Cell Res. 13(4): 251–256

Chesné, P., Adenot, P., Boulanger, L., and Renard, J. P. (2001) Somatic nuclear trasnfer in the rabbit. Theriogenology 55: 260 (abst).

Chesné, P., Adenot, P. G., Viglietta, C., Baratte, M., Boulanger, L., and Renard, J. P. (2002) Cloned rabbits produced by nuclear transfer from adult somatic cells. Nat. Biotechnol. 20(4): 366–369.

Chrenek, P., Vasicek, D., Makarevich, A. V., Jurcik, R., Suvegova, K., Parkanyi, V., Bauer, M., Rafay, J., Batorova, A., Paleyanda, R. K. (2005) Increased transgene integration efficiency upon microinjection of DNA into both pronuclei of rabbit embryos. Transgenic Res. 14(4): 417–28

Christians, E., Rao, V. H., and Renard, J. P. (1994) Sequential acquisition of transcriptional control during early embryonic development in the rabbit. Dev. Biol. 164: 160–172.

Cibelli, J. B., Stice, S. L., Golueke, P. J., Kane, J. J., Jerry, J., Blackwell, C., de Leon, F. A. P., and Robl, J. M. (1998) Cloned transgenic calves produced from nonquiescent fetal fibroblasts. Science 280: 1256–1258.

Collas, P. and Robl, J. M. (1990) Factors affecting the efficiency of nuclear transplantation in the rabbit embryo. Biol. Reprod. 43: 877–884.

Collas, P. and Robl, J. M. (1991) Relationship between nuclear remodeling and development in nuclear transplant rabbit embryos. Biol. Reprod. 45: 455–465.

Collas, P., Duby, R. T., and Robl, J. M. (1991) In vitro development of rabbit pronuclear embryos in rabbit peritoneal fluid. Biol. Reprod. 44: 1100–1107.

Collas, P., Balise, J. J., and Robl, J. M. (1992a) Influence of cell cycle stage of the donor nucleus on development of nuclear transplant rabbit embryos. Biol. Reprod. 46(3): 492–500.

Collas, P., Pinto-Correia, C., Ponce de Leon, F. A., and Robl, J. M. (1992b) Effect of donor cell cycle stage on chromatin and spindle morphology in nuclear transplant rabbit embryos. Biol. Reprod. 46(3): 501–511.

Deng, M. and Yang, X. J. (2001) Full term development of rabbit oocytes fertilized by Intracytoplasmic sperm injection. Mol. Reprod. Dev. 59: 38–43.

Denker, H. W. and Gerdes, H. J. (1979) The dynamic structure of rabbit blastocyst coverings. I. Transformation during regular preimplantation development. Anat. Embryol. (Berl) 157: 15–34.

Dinnyes, A., Dai, Y. P., Jiang, S., and Yang, X. Z. (2000) High developmental rates of vitrified bovine oocytes following parthenogenetic activation, in vitro fertilization, and somatic cell nuclear transfer. Biol. Reprod. 63: 513–518.

Dinnyes, A., Dai, Y., Barber, M., Liu, L., Xu, J., Zhou, P., and Yang, X. (2001a) Development of cloned embryos from adult rabbit fibroblasts: effect of activation treatment and donor cell preparation. Biol. Reprod. 64: 257–263.

Dinnyes, A., King, T., Wilmut, I., and De Sousa, P. A. (2001b) Sheep somatic cell nuclear transfer: effect of breed and culture system on embryonic and fetal development. Theriogenology 55: 264.

Dinnyes, A., Tian, X. C., and Yang, X. (2002) Cloning of rabbits in "Principles of cloning" (Jose Cibelli, Robert P. Lanza, Keith H.S. Campbell, Michael D. West, Eds.). Academic, San Diego, CA, pp. 343–366

Dinnyes, A., Tian, X. C., and Yang, X. (2008) Epigenetic Regulation of Foetal Development in Nuclear Transfer Animal Models. Reprod. Domes. Anim. 43: 302–309

Du, F., Giles, J. R., Foote, R. H., Graves, K. H., Yang, X., and Moreadith, R. W. (1995) Nuclear transfer of putative rabbit embryonic stem cells leads to normal blastocyst development. J. Reprod. Fertil. 104: 219–223.

Edwards, J. A. (1968) The external development of the rabbit and rat embryo. In "Advances in Teratology" (Woollam, D. H. M., Ed.), Academic, New York, pp. 239–262.

Ellington, J. E., Farrell, P. B., Simkin, M. E., Foote, R. H., Goldman, E. E., and McGrath, A. B. (1990) Development and survival after transfer of cow embryos cultured from 1–2-cells to morulae or blastocysts in rabbit oviducts or in a simple medium with bovine oviduct epithelial cells. J. Reprod. Fertil. 89: 293–299.

Enright, B. P., Kubota, C., Yang, X., and Tian, X. C. (2003) Epigenetic characteristics and development of embryos cloned from donor cells treated by trichostatin A or 5-aza-2′-deoxycytidine. Biol Reprod. 69: 896–901.

Escriba, M. J. and Garcia-Ximenez, F. (1999) Electroactivation of rabbit oocytes in an hypotonic pulsing medium and parthenogenetic in vitro development without cytochalasin B- diploidizing pretreatment. Theriogenology 51: 963–973.

Escriba, M. J. and Garcia-Ximenez, F. (2000) Influence of sequence duration and number of electrical pulses upon rabbit oocyte activation and parthenogenetic in vitro development. Anim. Reprod. Sci. 59: 99–107.

Evans, M. J and Kaufman, M. (1981) Establishment in culture of pluripotencial stem cells from mouse embryos. Nature 292: 151–156

Fang, Z. F., Gai, H., Huang, Y. Z., Li, S. G., Chen, X. J., Shi, J. J., Wu, L., Liu, A., Xu, P., and Sheng, H. Z. (2006) Rabbit embryonic stem cell lines derived from fertilized, parthenogenetic or somatic cell nuclear transfer embryos. Exp. Cell Res. 312(18): 3669–3682.

Farrell, P. B. and Foote, R. H. (1995) Beneficial effects of culturing rabbit zygotes to blastocysts in 5% oxygen and 10% carbon dioxide. J. Reprod. Fertil. 103: 127–130.

Feussner, E. L., Lightkep, G. E., Hennesy, R. A., Hoberman, A. M., and Christian, M. S. (1992) A decade of rabbit fertility data: study of historical control animals. Teratology 46: 349–365.

Fischer, B. (1989) Effects of asynchrony on rabbit blastocyst development. J. Reprod. Fertil. 86: 479–491.

Fischer, B., Mootz, U., Denker, H. W., Lambertz, M., and Beier, H. M. (1991) The dynamic structure of rabbit blastocyst coverings. III. Transformation of coverings under non-physiological developmental conditions. Anat. Embryol. (Berl) 183: 17–27.

Fissore, R. A. and Robl, J. M. (1993) Sperm, inositol trisphosphate, and thimerosal-induced intracellular Ca2 + elevations in rabbit eggs. Dev. Biol. 159: 122–130.

Fissore, R. A. and Robl, J. M. (1994) Mechanism of calcium oscillations in fertilized rabbit eggs. Dev. Biol. 166: 634–642.

Foote, R. H. and Carney, E. W. (2000) The rabbit as a model for reproductive and developmental toxicity studies. Reprod. Toxicol. 14: 477–493.

Franke, W. W., Schmid, E., Osborn, M., and Weber, K. (1978) Different intermediate-sized filaments distinguished by immunofluorescence microscopy. Proc. Natl. Acad. Sci. U. S. A 75: 5034–5038.

Franke, W. W., Schmid, E., Winter, S., Osborn, M., and Weber, K. (1979) Widespread occurrence of intermediate-sized filaments of the vimentin- type in cultured cells from diverse vertebrates. Exp. Cell. Res. 123: 25–46.

Galat, V. V., Lagutina, I. S., Mesina, M. N., Chernich, V. J., and Prokofiev, M. I. (1999) Developmental potential of rabbit nuclear transfer embryos derived from donor fetal fibroblast. Theriogenology 51: 203 (abst.)-203.

Giles, J. R. and Foote, R. H. (1997) Effects of gas atmosphere, platelet-derived growth factor and leukemia inhibitory factor on cell numbers of rabbit embryos cultured in a protein-free medium. Reprod. Nutr. Dev. 37: 97–104.

Giles, J. R., Yang, X., Mark, W., and Foote, R. H. (1993) Pluripotency of cultured rabbit inner cell mass cells detected by isozyme analysis and eye pigmentation of fetuses following injection into blastocysts or morulae. Mol. Reprod. Dev. 36: 130–138.

Graves, K. H. and Moreadith, R. W. (1993) Derivation and characterization of putative pluripotential embryonic stem cells from preimplantation rabbit embryos. Mol. Reprod. Dev. 36(4): 424–433.

Grocholova, R., Petr, J., Rozinek, J., and Jilek, F. (1997) The protein phosphatase inhibitor okadaic acid inhibits exit from metaphase II in parthenogenetically activated pig oocytes. J. Exp. Zool. 277: 49–56.

Hafez, E. S. E. (1964) Effects of over-crowding in utero on implantation and fetal development in the rabbit. J. Exp. Zool. 156: 269–288.

Hagen, K. W. (1974). Colony Husbandry. In "The Biology of the Laboratory Rabbit" (Weisbroth, S. H., Flatt, R. E., and Kraus, A. L., Eds.), Academic, New York, pp. 23–47.
Hammer, R. E., Pursel, V. G., Rexroad, C. E., Wall, R. J., Bolt, D. J., Ebert, K. M., Palmiter, R. D., and Brinster, R. L. (1985) Production of transgenic rabbits, sheep and pigs by microinjection. Nature 315: 680–683.
Harkness, J. E. and Wagner, J. E. (1995) The Biology and Medicine of Rabbits and Rodents. William and Wilkins, Baltimore, MD.
Heape, W. (1890) Preliminary note on the transplantation and growth of mammalian ova within a uterine foster-mother. Proc. R. Soc. Lond., B. Biol. Sci. 48: 457–459.
Henrion, G., Renard, J. P., Chesne, P., Oudin, J. F., Maniey, D., Brunet, A., Osborne, H. B., and Duranthon, V. (2000) Differential regulation of the translation and the stability of two maternal transcripts in preimplantation rabbit embryos. Mol. Reprod. Dev. 56: 12–25.
Heyman, Y. and Renard, J. P. (1996) Cloning of domestic species. Anim. Reprod. Sci. 42: 427–436.
Heyman, Y., Chesne, P., and Renard, J. P. (1990) Reprogrammation complete de noyaux embryonnaires congeles apres transfert nucleaire chez le lapin. C. R. Acad. Sci. Paris 311: 321–326.
Hornsby, P. J, Yang, L., Gunter, L. E. (1992) Demethylation of satellite I DNA during senescence of bovine adrenocortical cells in culture. Mutat. Res. 275:13–19.
Iannaccone, P. M., Taborn, G. U., Garton, R. L, Caplice, M. D., and Brenin, D. R. (1994) Pluripotent embryonic stem cells from the rat ae capable of producing chimeras. Dev. Biol. 163: 288–292.
Inoue, K., Ogonuki, N., Yamamoto, Y., Noguchi, Y., Takeiri, S., Nakata, K, Miki, H., Kurome, M., Nagashima, H., and Ogura, A. (2002) Improved postimplantation development of rabbit nuclear transfer embryos by activation with inositol 1,4,5-trisphosphate. Cloning Stem Cells 4(4): 311–317.
Jiang, Y., Chen, T., Nan, C. L., Ouyang, Y. C., Sun, Q. Y., and Chen, D. Y. (2005) In vitro culture and mtDNA fate of ibex-rabbit nuclear transfer embryos. Zygote 13(3): 233–240.
Jin, D. I., Kim, D. K., Im, K. S., and Choi, W. S. (2000) Successful pregnancy after transfer of rabbit blastocysts grown in vitro from single-cell zygotes. Theriogenology 54: 1109–1116.
Joung, S. Y., Kim, H. J., Choi, W. S., Im, K. S., Lee, S. H., Park, C. S., and Jin, D. I. (2004) Effects of transferring in vitro-cultured rabbit embryos to recipient oviducts on mucin coat deposition, implantation and development. Zygote 12(3): 215–219.
Kane, M. T. and Foote, R. H. (1971) Factors affecting blastocyst expansion of rabbit zygotes and young embryos in defined media. Biol. Reprod. 4: 41–47.
Kanka, J., Hozak, P., Heyman, Y., Chesne, P., Degrolard, J., Renard, J. P., and Flechon, J. E. (1996) Transcriptional activity and nucleolar ultrastructure of embryonic rabbit nuclei after transplantation to enucleated oocytes. Mol. Reprod. Dev. 43: 135–144.
Kasai, M., Hamaguchi, Y., Zhu, S. E., Miyake, T., Sakurai, T., and Machida, T. (1992) High survival of rabbit morulae after vitrification in an ethylene glycol-based solution by a simple method. Biol. Reprod. 46: 1042–1046.
Kasinathan, P., Knott, J. G., Moreira, P. N., Burnside, A. S., Joseph, J. D., and Robl, J. M. (2001) Effect of fibroblast donor cell age and cell cycle on development of bovine nuclear transfer embryos in vitro. Biol. Reprod. 64: 1487–1493.
Kato, Y., Tani, T., Sotomaru, Y., Kurokawa, K., Kato, J., Doguchi, H., Yasue, H., and Tsunoda, Y. (1998) Eight calves cloned from somatic cells of a single adult. Science 282: 2095–2098.
Kato, Y., Yabuuchi, A., Motosugi, N., Kato, J., and Tsunoda, Y. (1999) Developmental potential of mouse follicular epithelial cells and cumulus cells after nuclear transfer. Biol. Reprod. 61: 1110–1114.
Kauffman, R. D., Schmidt, P. M., Rall, W. F., and Hoeg, J. M. (1998).Superovulation of rabbits with FSH alters in vivo development of vitrified morulae. Theriogenology 50: 1081–1092.
Kennelly, J. J. and Foote, R. H. (1965) Superovulatory response of pre- and post-puberal rabbits to commercially available gonadotrophins. J. Reprod. Fertil. 9: 131–145.
Kidder, J. D., Roberts, P. J., Simkin, M. E., Foote, R. H., and Richmond, M. E. (1999) Nonsurgical collection and nonsurgical transfer of preimplantation embryos in the domestic rabbit

(Oryctolagus cuniculus) and domestic ferret (Mustela putorius furo). J. Reprod. Fertil. 116: 235–242.
Kishigami, S., Mizutani, E., Ohta, H., Hikichi, T., Thuan, N. V., Wakayama, S., Bui, H. T., and Wakayama, T. (2006) Significant improvement of mouse cloning technique by treatment with trichostatin A after somatic nuclear transfer. Biochem Biophys Res Commun. 340(1): 183–9.
Koyama, H., Suzuki, H., Yang, X., Jiang, S., Foote, R. H. (1994) Analysis of polarity of bovine and rabbit embryos by scanning electron microscopy. Biol. Reprod. 50: 163–70.
Kubota, C., Yamakuchi, H., Todoroki, J., Mizoshita, K., Tabara, N., Barber, M., and Yang, X. Z. (2000) Six cloned calves produced from adult fibroblast cells after long-term culture. Proc. Natl. Acad. Sci. U. S. A 97: 990–995.
Lagutina, I. S., Zakhartchenko, V. I., and Prokofiev, M. I. (1999) Nuclear transfer in rabbits and factors affecting it efficiency. Theriogenology 51: 207 (abst).
Lagutina, I. S., Mezina, M. N., Chernikh, V. J., Prokofiev, M. I., and Galat, V. V. (2000) Developmental potential of rabbit nuclear transfer embryos produced by various fusion/activation protocols. Theriogenology 53: 230 (abst).
Lanza, R. P., Cibelli, J. B., Diaz, F., et al. (2000) Cloning of an endangered species (Bos gaurus) using interspecies nuclear transfer. Cloning 2: 79–91.
Lebas, F. (1970) Description d'une machine a traire les lapines. Ann Zootech 19: 223–228.
Leiser, R. and Denker, H. W. (1988) The dynamic structure of rabbit blastocyst coverings. II. Ultrastructural evidence for a role of the trophoblast in neozona formation. Anat. Embryol. (Berl) 179: 129–134.
Li, G. P., Chen, D. Y., Lian, L., Han, Z. M., Zhu, Z. Y., and Seidel, G. E. Jr. (2002) Rabbit cloning: improved fusion rates using cytochalasin B in the fusion buffer. Mol. Reprod. Dev. 61(2): 187–191.
Li, J. and Foote, R. H. (1996) Differential sensitivity of one-cell and two-cell rabbit embryos to sodium chloride and total osmolarity during culture into blastocysts. J. Reprod. Fertil. 108: 307–312.
Li, J., Foote, R. H., and Simkin, M. (1993) Development of rabbit zygotes cultured in protein-free medium with catalase, taurine, or superoxide dismutase. Biol. Reprod. 49: 33–37.
Li, S., Chen, X., Fang, Z., Shi, J., and Sheng, H. Z. (2006) Rabbits generated from fibroblasts through nuclear transfer. Reproduction 131(6): 1085–1090.
Liu, J. L., Sung, L. Y., Du, F., Julian, M., Jiang, S., Barber, M., Xu, J., Tian, X. C., and Yang, X. (2004) Differential development of rabbit embryos derived from parthenogenesis and nuclear transfer. Mol. Reprod. Dev. 68(1): 58–64.
Liu, L., Ju, J. C., and Yang, X. (1998) Differential inactivation of maturation-promoting factor and mitogen-activated protein kinase following parthenogenetic activation of bovine oocytes. Biol. Reprod. 59: 537–545.
Lonergan, P., Dinnyes, A., Fair, T., Yang, X., and Boland, M. (2000) Bovine oocyte and embryo development following meiotic inhibition with butyrolactone I. Mol. Reprod. Dev. 57: 204–209.
Meng, Q., Polgar, Z., Liu, J., and Dinnyes, A. (2008) Effect of Trichostatine A treatment on the term development of somatic cell nuclear transfer rabbit embryos Reprod. Fertil. Dev. 20(1): 103
Marcus, G. E., Shum, F. T., and Goldman, S. L. (1990) A device for collecting milk from rabbits. Lab. Anim. Sci. 40: 219–221.
Maurer, R. R., Hunt, W. L., Van Vleck, L. D., and Foote, R. H. (1968) Developmental potential of superovulated rabbit ova. J. Reprod. Fertil. 15: 171–175.
Maurer, R. R. (1978) Advances in Rabbit Embryo Culture. In "Methods in Mammalian Reproduction" (Daniel, J. C., Ed.), Academic, New York, pp. 259–272.
Mitalipov, S. M., White, K. L., Farrar, V. R., Morrey, J., and Reed, W. A. (1999) Development of nuclear transfer and parthenogenetic rabbit embryos activated with inositol 1,4,5-trisphosphate. Biol. Reprod. 60: 821–827.
Modlinski, J. A. and Smorag, Z. (1991) Preimplantation development of rabbit embryos after trasnfer of embryonic nuclei into different cytoplasmic environment. Mol. Reprod. Dev. 28: 361–372.

Moens, A., Chastant, S., Chesne, P., Flechon, J. E., Betteridge, K. J., and Renard, J. P. (1996) Differential ability of male and female rabbit fetal germ cell nuclei to be reprogrammed by nuclear transfer. Differentiation 60: 339–345.
Murakami, H. and Imai, H. (1996) Successful implantation of in vitro cultured rabbit embryos after uterine transfer: a role for mucin. Mol. Reprod. Dev. 43: 167–170.
Nussbaum, D. J. and Prather, R. S. (1995) Differential effects of protein synthesis inhibitors on porcine oocyte activation. Mol. Reprod. Dev. 41: 70–75.
Onishi, A., Iwamoto, M., Akita, T., Mikawa, S., Takeda, K., Awata, T., Hanada, H., and Perry, A. C. F. (2000) Pig cloning by microinjection of fetal fibroblast nuclei. Science 289: 1188–1190.
Orsini, M. W. (1962) Study of ovo-implantation in the hamster, rat, mouse, guinea-pig, and rabbit in cleared uterine tracts. J. Reprod. Fertil. 3: 288–293.
Ozil, J. P. (1990) The parthenogenetic development of rabbit oocytes after repetitive pulsatile electrical stimulation. Development 109: 117–127.
Ozil, J. P. and Huneau, D. (2001) Activation of rabbit oocytes: the impact of the Ca2 + signal regime on development. Development 128: 917–928.
Park, C. S., Jeon, B. G., Lee, K. M., Yin, X. J., Cho, S. K., Kong, I. K., Lee, H. J., and Choe, S. Y. (1998) Production of cloned rabbit embryos and offsprings by nuclear transplantation using in vitro matured oocytes. Theriogenology 49: 325 (abst).
Petters, R. M. and Wells, K. D. (1993) Culture of pig embryos. J. Reprod. Fertil. Suppl. 48: 61–73.
Pinto-Correia, C., Long, C. R., Chang, T., and Robl, J. M. (1995) Factors involved in nuclear reprogramming during early development in the rabbit. Mol. Reprod. Dev. 40: 292–304
Piotrowska, K., Modlinski, J. A., Korwin-Kossakowski, M., and Karasiewicz, J. (2000) Effects of preactivation of ooplasts or synchronization of blastomere nuclei in G1 on preimplantation development of rabbit serial nuclear transfer embryos. Biol. Reprod. 63: 677–682.
Polejaeva, I. A., Chen, S. H., Vaught, T. D., Page, R. L., Mullins, J., Ball, S., Dai, Y. F., Boone, J., Walker, S., Ayares, D. L., Colman, A., and Campbell, K. H. S. (2000) Cloned pigs produced by nuclear transfer from adult somatic cells. Nature 407: 86–90.
Prather, R. S., Mayes, M. A., and Murphy, C. N. (1997) Parthenogenetic activation of pig eggs by exposure to protein kinase inhibitors. Reprod. Fertil. Dev. 9: 539–544.
Rao, V. H., Heyman, Y., Chesne, P., and Renard, J. P. (1998) Freezing and recloning of nuclear transfer embryos in rabbit. Theriogenology 49: 328 (abst).
Renard, J. P., Bui, X. N., and Garnier, V. (1984) Two-step freezing of two-cell rabbit embryos after partial dehydration at room temperature. J. Reprod. Fertil. 71: 573–580.
Rideout, W. M., Wakayama, T., Wutz, A., Eggan, K., Jackson-Grusby, L., Dausman, J., Yanagimachi, R., and Jaenisch, R. (2000) Generation of mice from wild-type and targeted ES cells by nuclear cloning. Nat Genet. 24(2): 109–110.
Rybouchkin, A., Kato, Y., and Tsunoda, Y. (2006) Role of histone acetylation in reprogramming of somatic nuclei following nuclear transfer. Biol Reprod. 74: 1083–1089
Santos, F. and Dean, W. (2004) Epigenetic reprogramming during early development in mammals. Reproduction 127: 643–651.
Schnieke, A. E., Kind, A. J., Ritchie, W. A., Mycock, K., Scott, A. R., Ritchie, M., Wilmut, I., Colman, A., and Campbell, K. H. S. (1997) Human factor IX transgenic sheep produced by transfer of nuclei from transfected fetal fibroblasts. Science 278: 2130–2133.
Schoonjans, L., Albright, G. M., Li, J. L., Collen, D., and Moreadith, R. W. (1996) Pluripotential rabbit embryonic (ES) cells are capable of forming over coat color chimeras following injection into the blastocyst. Mol. Reprod. 45:439–443.
Schultz, R. M., Davis, W., Jr., Stein, P., and Svoboda, P. (1999) Reprogramming of gene expression during preimplantation development. J. Exp. Zool. 285: 276–282.
Shi, L, Miao, Y., Ouyang, Y., Huang, J., Lei, Z., Yang, J., Han, Z., Son, X., Sun, Q., and Chen, D. (2008a) Trichostatin A (TSA) improves the development of rabbit-rabbit intraspecies cloned embryos, but not rabbit-human interspecies cloned embryos. Dev Dyn. 273: 640–648.
Shi, L, Ai, J., Ouyang, Y., Huang, J., Lei, Z., Wang, Q., Yin, S., Han, Z., Sun, Q., and Chen, D. (2008b) Trichostatin A and nuclear reprogramming of cloned rabbit embryos. J Anim Sci. 86: 1106–1113.

Shi, W., Dirim, F., Wolf, E., Zakhartchenko, V., and Haaf, T. (2004) Methylation reprogramming and chromosomal aneuploidy in in vivo fertilized and cloned rabbit preimplantation embryos. Biol. Reprod. 71(1): 340–347.
Shiga, K., Fujita, T., Hirose, K., Sasae, Y., and Nagai, T. (1999) Production of calves by transfer of nuclei from cultured somatic cells obtained from Japanese black bulls. Theriogenology 52: 527–535.
Smith, L. C. and Wilmut, I. (1989) Influence of nuclear and cytoplasmic activity on the development in vivo of sheep embryos after nuclear transplantation. Biol. Reprod. 40: 1027–1035.
Stice, S. L. and Keefer, C. L. (1993) Multiple generational bovine embryo cloning. Biol. Reprod. 48: 715–719.
Stice, S. L. and Robl, J. M. (1988) Nuclear reprogramming in nuclear transplant rabbit embryos. Biol. Reprod. 39: 657–664.
Stice, S. L. and Robl, J. M. (1990) Activation of mammalian oocytes by a factor obtained from rabbit sperm. Mol. Reprod. Dev. 25: 272–280.
Stice, S. L., Keefer, C. L., Maki-Laurila, M., and Matthews, L. (1993) Donor blastomere cell cycle stage affects developmental competence of bovine nuclear transfer embryos. Theriogenology 39: 318 (abst).
Susko-Parrish, J. L., Leibfried-Rutledge, M. L., Northey, D. L., Schutzkus, V., and First, N. L. (1994) Inhibition of protein kinases after an induced calcium transient causes transition of bovine oocytes to embryonic cycles without meiotic completion. Dev. Biol. 166: 729–739.
Swann, K. and Lai, F. A. (1997) A novel signalling mechanism for generating Ca2 + oscillations at fertilization in mammals. Bioessays 19: 371–378.
Tan, J. H., Zhou, Q., Li, Z. Y., Sun, X. S., Liu, Z. H., and He, G. X. (1997) Embryo cloning by nuclear transplantation in rabbits. Theriogenology 47: 237 (abst).
Telford, N. A., Watson, A. J., and Schultz, G. A. (1990) Transition from maternal to embryonic control in early mammalian development: a comparison of several species. Mol. Reprod. Dev. 26: 90–100.
Thompson, E. M., Legouy, E., Christians, E., and Renard, J. P. (1995) Progressive maturation of chromatin structure regulates HSP70.1 gene expression in the preimplantation mouse embryo. Development 121: 3425–3437.
Thomson, J. A., Kalishman, J., Golos, T. G., Durning, M., Harris, C. P., Becker, R. A., and Hearn, J. P. (1995) Isolation of a primate embryonic stem cell line. Proc. Natl.Acad. Sci. USA 92: 7844–7848.
Thomson, J. A., Itskovitz-Eldor, J., Shapiro, S. S., Waknitz, M. A., Swiergiel, J. J., Marshal, Swiergiel, V. S., and Jones, J. M. (1998) Embryonc stem cell lines derived from human blastocyst. Science 282: 1145–1147.
Totey, S. M., Singh, G., Taneja, M., Pawshe, C. H., and Talwar, G. P. (1992) In vitro maturation, fertilization and development of follicular oocytes from buffalo (Bubalus bubalis). J. Reprod. Fertil. 95: 597–607.
Tsunoda, Y., Soma, T., and Sugie, T. (1982) Effect of post-ovulatory age of recipient on survival of frozen-thawed rabbit morulae. J. Reprod. Fertil. 65: 483–487.
Vajta, G. and Gjerris, M. (2006) Science and technology of farm animal cloning: State of the art. Animal Reprod. Sci. 92: 211–230.
Varian, N. B., Maurer, R. R., and Foote, R. H. (1967) Ovarian response and cleavage rate of ova in control and FSH-primed rabbits receiving varying levels of luteinizing hormone. J. Reprod. Fertil. 13: 67–73.
Vignon, X., Chesne, P., Le Bourhis, D., Flechon, J. E., Heyman, Y., and Renard, J. P. (1998) Developmental potential of bovine embryos reconstructed from enucleated matured oocytes fused with cultured somatic cells. C. R. Acad. Sci. Paris 321: 735–745.
Vincent, C., Garnier, V., Heyman, Y., and Renard, J. P. (1989) Solvent effects on cytoskeletal organization and in-vivo survival after freezing of rabbit oocytes. J. Reprod. Fertil. 87: 809–820.
Wakayama, T. and Yanagimachi, R. (2001) Mouse cloning with nucleus donor cells of different age and type. Mol. Reprod. Dev. 58: 376–383.
Wakayama, T., Perry, A. C. F., Zuccotti, M., Johnson, K. R., and Yanagimachi, R. (1998) Full-term development of mice from enucleated oocytes injected with cumulus cell nuclei. Nature 394: 369–374.

Wang, F, Kou, Z., Zhang, Y., and Gao, S. (2007) Dynamic Reprogramming of Histone Acetylation and Methylation in the First Cell Cycle of Cloned Mouse Embryos. Biol Reprod. 77: 1007–1016.

Wall, R. J., Powell, A. M., Paape, M. J., Kerr, D. E., Bannermann, D. D., Pursel, V. G., Wells, K. D., Talbot, N., and Hawk, H. W. (2005) Genetically enhanced cows resist intramammary Staphylococcus aureus infection. Nat. Biotechnol. 23: 445–451.

Wang, S., Tang, X., Niu, Y., Chen, H., Li, B., Li, T., Zhang, X., Hu, Z., Zhou, Q., Ji, W. (2006) Generation and characterization of rabbit embryonic stem cells. Stem Cells 25(2): 481–489.

Wang, W., Sun, Q., Hosoe, M., and Shioya, Y. (1997) Calcium- and meiotic-spindle-independent activation of pig oocytes by the inhibition of staurosporine-sensitive protein kinases. Zygote 5: 75–82.

Wells, D. N., Misica, P. M., McMillan, W. H., and Tervit, H. R. (1998) Production of cloned bovine fetuses following nuclear transfer using cells from a fetal fibroblast cell line. Theriogenology 49: 330.

Willadsen, S. M. (1986) Nuclear transplantation in sheep embryos. Nature 320: 63–65.

Willadsen, S. M. (1989) Cloning of sheep and cow embryos. Genome 31: 956–962.

Wilmut, I., Schnieke, A. E., McWhir, J., Kind, A. J., and Campbell, K. H. S. (1997) Viable offspring derived from fetal and adult mammalian cells. Nature 385: 810–813.

Xu, J., Sung, L.-Y., Zhang, J., Tian, X., Chen, Y. E., Yang, X., and Du, F. (2007) Trichostatin A Improved the Quality of Rabbit Nuclear Transfer Embryos. Reprod Fertil Dev. 19: 165.

Yang, F., Hao. R., Kessler, B., Brem, G., Wolf, E., and Zakhartchenko, V. (2007) Rabbit somatic cell cloning: effects of donor cell type, histone acetylation status and chimeric embryo complementation. Reproduction 133(1): 219–230.

Yang, X., Chen, Y., Chen, J., and Foote, R. H. (1990a) Potential of hypertonic medium treatment for embryo micromanipulation: I. Survival of rabbit embryos in vitro and in vivo following sucrose treatment. Mol. Reprod. Dev. 27: 110–117.

Yang, X., Zhang, L., Kovacs, A., Tobback, C., and Foote, R. H. (1990b) Potential of hypertonic medium treatment for embryo micromanipulation: II. Assessment of nuclear transplantation methodology, isolation, subzona insertion, and electrofusion of blastomeres to intact or functionally enucleated oocytes in rabbits. Mol. Reprod. Dev. 27: 118–129.

Yang, X. (1991) Featured article: Embryo cloning by nuclear transfer in cattle and rabbits. Embryo Transfer Newslett. 9: 10–22.

Yang, X., Jiang, S., Kovacs, A., and Foote, R. H. (1992) Nuclear totipotency of cultured rabbit morulae to support full-term development following nuclear transfer. Biol. Reprod. 47: 636–643.

Yang, X. Z. and Foote, R. H. (1987) Production of identical twin rabbits by micromanipulation of embryos. Biol. Reprod. 37: 1007–1014.

Yang, X. Z. and Foote, R. H. (1990) Survival of bisected rabbit morulae transferred to synchronous and asynchronous recipients. Mol. Reprod. Dev. 26: 6–11.

Yin, X. J., Tani, T., Kato, Y., and Tsunoda, Y. (2000) Development of rabbit parthenogenetic oocytes and nuclear-transferred oocytes receiving cultured cumulus cells. Theriogenology 54: 1469–1476.

Yin, X. J., Kato, Y., and Tsunoda, Y. (2002a) Effect of enucleation procedures and maturation conditions on the development of nuclear-transferred rabbit oocytes receiving male fibroblast cells. Reproduction 124(1): 41–47.

Yin, X. J., Kato, Y., and Tsunoda, Y. (2002b) Effect of delayed enucleation on the developmental potential of nuclear-transferred oocytes receiving adult and fetal fibroblast cells. Zygote 10(3): 217–222.

Zakhartchenko, V., Alberio, R., Stojkovic, M., Prelle, K., Schernthaner, W., Stojkovic, P., Wenigerkind, H., Wanke, R., Duchler, M., Steinborn, R., Mueller, M., Brem, G., and Wolf, E. (1999) Adult cloning in cattle: Potential of nuclei from a permanent cell line and from primary cultures. Mol. Reprod. Dev. 54: 264–272.

Zhou, Q., Li, Z. Y., Liu, Z. H., Sun, X. S., He, G. X., and Tan, J. H. (1997) Nuclear transplantation by microinjection in rabbits. Theriogenology 47: 239 (abst).

Ziomek, C. A., Chatot, C. L., and Manes, C. (1990) Polarization of blastomeres in the cleaving rabbit embryo. J. Exp. Zool. 256: 84–91.

Chapter 11
The European Rabbit Genome

Jessica Alföldi, Federica Di Palma, and Kerstin Lindblad-Toh

The rabbit genome was chosen to be sequenced to low coverage (twofold) in an effort to annotate the human genome using 24 low coverage mammalian genomes. The European rabbit was among a first group of eight mammals selected for this sequencing effort for identification of conserved and functional elements in the human genome. The individual chosen to be sequenced was a female inbred Thorbecke New Zealand White rabbit. The rabbit's twofold assembly was completed in 2005, yielding an N50 contig size of 3.3 kb and an N50 scaffold size of 55 kb.

While a low coverage assembly is a good resource for the rabbit research community, it cannot provide the same level of information as a high coverage assembly. And so in 2006, the rabbit was chosen to be sequenced to full (sevenfold) coverage at the Broad Institute based on its importance to biomedical research and to aid in the reconstruction of the ancestral placental mammal genome. This new high coverage rabbit assembly will provide better information about orthologous genes, to help those studying disease models and holding pre-clinical drug trials. The full sequence of the rabbit's MHC region will aid immunological research. This genome will also enable rabbit transgenics, gene expression and gene therapy experiments. And it will serve as a basis for the study of rabbit population diversity, speciation and domestication. The Broad Institute expects the sevenfold rabbit genome assembly to be completed by the end of 2008.

Further information on the rabbit genome project at the Broad Institute can be found at: http://www.broad.mit.edu/node/709.

The entire twofold rabbit genome assembly can be found at: http://www.broad.mit.edu/ftp/pub/assemblies/mammals/rabbit/.

To further explore the rabbit genome, we recommend that you visit the rabbit webpages at NCBI and Ensembl:

http://www.ncbi.nlm.nih.gov/projects/genome/guide/rabbit/

http://www.ensembl.org/Oryctolagus_cuniculus/index.html

J. Alföldi (✉)
Broad Institute, Cambridge, MA 02141, USA
e-mail: kersli@broad.mit.edu

L.-M. Houdebine and J. Fan (eds.), *Rabbit Biotechnology: Rabbit Genomics, Transgenesis, Cloning and Models*,

Index

FSC
www.fsc.org
MIX
Papier | Fördert gute Waldnutzung
FSC® C083411

Kolibri 360 GmbH
Ferdinand-Jühlke-Straße 7
99095 Erfurt, Deutschland
produktsicherheit@kolibri360.de